Design and Nature: A Partnership

Edited by Kate Fletcher, Louise St. Pierre and Mathilda Tham

Organised as a dialogue between nature and design, this book explores design ideas, opportunities, visions and practices through relating and uncovering experience of the natural world.

Presented as an edited collection of 25 wide-ranging short chapters, the book explores the possibility of new relations between design and nature, beyond human mastery and understandings of nature as resource and by calling into question the longstanding role for design as agent of capitalism. The book puts forward ways in which design can form partnerships with living species and examines designers' capacities for direct experience, awe, integrated relationships and new ways of knowing. It covers:

· New design ethics of care

· Indigenous perspectives

· Prototyping with nature

· Methods for new design and nature relations

· A history of design and nature

· Animist beliefs

· De-centering human-centered design

· Understanding nature has power and agency

Design and Nature: A Partnership is a rich resource for designers who wish to learn to engage with sustainability from the ground up.

Kate Fletcher is a Professor at the Centre of Sustainable Fashion, University of the Arts London, UK, where she explores design for sustainability.

Louise St. Pierre is an Associate Professor at Emily Carr University, Canada, where she researches sustainability, contemplative practice and design's relationship with nature.

Mathilda Tham's work sits in a creative, feminist, activist space between design, futures studies and sustainability. She is Professor of Design, Linnaeus University, Sweden.

Design and Nature: A *Partnership*

Edited by ***Kate Fletcher, Louise St. Pierre*** *and* ***Mathilda Tham***

Routledge
Taylor & Francis Group
LONDON AND NEW YORK

earthscan
from Routledge

First published 2019
by Routledge
2 Park Square, Milton Park, Abingdon, Oxon OX14 4RN

and by Routledge
52 Vanderbilt Avenue, New York, NY 10017

Routledge is an imprint of the Taylor & Francis Group, an informa business

British Library Cataloguing-in-Publication Data
A catalogue record for this book is available from the British Library

Library of Congress Cataloging-in-Publication Data
A catalog record has been requested for this book

ISBN: 978-0-8153-6273-9 (hbk)
ISBN: 978-0-8153-6274-6 (pbk)
ISBN: 978-1-351-11151-5 (ebk)

Design by Katherine Gillieson
Cover art by Oscar Gillieson-Ashdown
Typeset in Goudy Old Style and Akzidenz Grotesk

Publisher's Note
This book has been prepared from camera-ready copy provided by Kate Fletcher, Louise St. Pierre and Mathilda Tham.

For Jude and Cole (KF)

For Rebecca (LSP)

For Paloma and Rosa Lulu (MT)

Contents

SECTION III: STANDING

SECTION IV: WALKING

Acknowledgements

This book is in your hands courtesy of the brightness and generosity of its contributing authors. We are hugely grateful to them for their tenacity in sticking with us through a lengthy writing and editing process and their imagination and experience for compiling a collection of compelling chapters.

Thank you to Avery Shaw and Michal Cabaj who have kept us lean and trim and organised. Superlative work, thank you both. We gratefully acknowledge that the Internal Research Fund at Emily Carr University supported our bringing you into this project.

Superlative thanks also to Katherine Gillieson for the book design. She was a steady eye and hand. We needed both. We love what you created.

From Louise, a thank you to my longtime reader Catherine Fraser. Dear friend, you helped me gather my thoughts and laugh about my sentences.

Nature has, in different ways, saved this book's editors over the last decade. We live in gratitude for the friendship and the collaboration that this partnership has brought us.

One of the first in-person meetings that we had about the book took place on a ferry heading to Vancouver Island off the south west coast of Canada. The sky was overcast that day, but the sea and the prospect of whales tempered everything with light.

Introduction

Ethos and agenda

This book is about searching for new relationships between design and nature. The three of us editors imagine these relationships may celebrate the diversity, abundance, wonder, rich histories and creativity of the world that humans share with their cospecies—and ensure the survival and continuity of all this life. In order to uncover these relationships, the book proposes a new orientation for design. Design is a practice that historically, at least in the Global North, has been aligned with progress, industry, capitalism and modernity. We contend that there is an alternate alliance for design: a partnership with nature. This partnership with nature disrupts dominant expectations of design, as it is incompatible with the drive of the modern economy. As Klein states, "Our economy is at war with many forms of life on earth, including human life" (2014: 21). The growth of industry, economies and production has paralleled the deterioration of rivers, air and global temperatures. Current assessments of the damaged state of Earth's systems (Steffen et al., 2015) bring about an almost paralysing sense of ecological urgency. A new name, The Anthropocene, indicating that humans are now driving changes to vital Earth systems, is proposed for this next epoch (Crutzen et al., 2007). Other proposed names include Capitalocene, Plasticene and Chthulucene (implying the ongoing dynamic interplay of forces) (Haraway, 2015). What is significant to us is that a new word is required at all. The call for a new word marks the extent of the tremendous geophysical change on Earth, the end of 10,000 years of the previous epoch, the Holocene. The Holocene was a climatically stable era which allowed humans to settle, pursue systematic agriculture, make dwellings and develop tools, and much more, including many activities which may be termed design. This stability is now no more. It is a frightening time.

This unpredictability of our home world challenges designers to invest in doing right by the Earth, an endeavour that is more radical than we, as a design community, had perhaps at first anticipated. We now know that ecological urgency demands more than a tweak of designed objects, design processes and even design systems. In fact, ecological urgency impels us to push against the forces that have birthed design, have shaped design for decades, and now confine it. Design is an activity that makes ideas, artefacts and spaces become real; design has tremendous agency, a capacity to change the way we live in the world. Yet design's capacity is typically harnessed by the corporate, political and industrial world to serve its goals (Boehnert, 2018). We can now see clearly that these goals have little to do with supporting life.

Looking at the history of design's engagement with nature, we see that it includes attempts to compromise, to bridge the needs of the Earth with the needs of production and economics. However, typically these compromises have closed down learning about the natural world before they have had a chance to develop into "new relationships, languages and commensurate practices that will only arise slowly out of actual engagement and new ways of being present to, and interacting with the [natural] world" (Jickling et al., 2018: 36). In order to uncover these new design practices, we need to allow ourselves the time and space to immerse in the grit of the natural world and to leave the 'safety' of some comfort zones in terms of how to know, be and relate. Changing design requires active and humble engagement with the natural world. We imagine a form of exploratory questioning that presumes no power and seeks no answers; looking only for relationships.

This book moves past the distractions of academic dispute, to advance discussions of nature as intellectually serious territory in their own right. We suggest that nature can be discussed on her own terms. This entails a dynamic, unpredictable and messy process rather than one that is delimited or framed by an academic discourse that attempts to pin down, claim or quantify knowledge and ways of knowing, and thereby instrumentalise knowledge. Traditional forms of intellectualism are a hindrance in the distance and power hierarchies they set up. New epistemologies of personal, experiential and subjective narratives open potential for ways of knowing that are not logical, or more than logical, and for practices other than those based in scientific methodology. We point to decoloniality, post-humanism and the resurgence of animism as a decentring of the human, peeling away the underlying elitism in the humanist tradition that will allow new narratives for design and nature; a nature that has agency and is a partner in our own re-learning. These positions are congruent with inclusive feminist perspectives, and a diverse distribution of responsibilities and care (such as proposed by Haraway, 2016 and Puig de la Bellacasa, 2017). They pose difficult challenges to the designer and to design, since such relational ontologies and epistemologies also mean the end of some modernist design icons as characterised by the star designer, the lone genius, the perfect solution or product, the finished building.

Partnering with nature therefore is a more disruptive and deeper change for design than many imagine when they read the friendly phrase "design and nature". Caught as we are in our industrial Modern paradigm, embedded in thousands of years of boundary making (Jickling, et al. 2018), to move design from its human centrings and to begin to imagine what Plumwood (2009) calls positive and multiple centrings is radical, heresy almost. While proclaiming primary affiliation with life on Earth should not, it seems to us, be contentious, this book inevitably seeds dissent with modernity, and with the roots of design. This is what a new alliance and partnership does. We hope that you, together with us, can accept the discomfort that venturing into new relationships can cause and trust that some letting go will be worth it. It is important to remind ourselves that many of the systems we live by are humanmade. We can therefore, with nature, make new, renew old; make better. Not only that but many of the design systems we live by have been and are heavily promoted and supported by very powerful agents of growth and modernity. Therefore, forging other pathways can feel counter-culture, unfamiliar even dangerous. We are confident, and as you will see—as are many of the contributing authors to this book—that there are other ways of being, doing, seeing. Of course, they won't be perfect—and this is good! These other pathways will open to us and grow if we give them our attention and trust.

How this book is organised

This book is made up of 24 short chapters, from design researchers and practitioners around the world, organised in four sections. It has an additional 'floating' essay that we have chosen to insert in the middle of the book because it gets to the heart of the matter of nature and design. This essay provides an exploration of the background and context of design in an expanded engagement with the more-than-human world and provides essential grounding for the four sections and the 24 chapter contributions.

The four sections are named after four postures. The postures—lying, sitting, standing and walking—make manifest basic modes of action and awareness that we see as essential to being present in the world and to participating in the work of design and nature. For us, each of the postures has a specific flavour and effect on human experience. They call forth a different way of seeing the world, and we have clustered the book's various contributions into groups accordingly with each posture and something of its thematic influence introduced at the beginning of each section. Our decision to structure the book around different types of engagement and action was encouraged by our interest in direct experience of the world, to which our fate is irrevocably tied. They are also the four postures of Buddhism. One Buddhist text, *Visuddhimagga* (Buddhaghosa, 2011: 124) makes the suggestion that practitioners should try different postures for three days each and then decide on the one that is most comfortable for them. We agree: we suggest that ideas and action around design and nature are tested out in different contexts where they are examined and explored against the backdrop of specific conditions.

It is worth saying that the process of assigning chapters to the various sections was far from an exact operation. Energy runs through all of the book's posture-sections and it also sparks across places between the sections. As such we tried to take a flowing approach to allocating contributions to the various sections, acknowledging that a good number of them could have easily been placed elsewhere and at the points where different types of action and awareness meet.

A note on writing style

As editors, we encouraged all of the book's contributors to write in the first person and to draw from direct experience. We felt that this was important because a first person narrative personalises content and insights, bringing them home in a way that more passive language structures often fail to do. It firmly underscores the intimacy of this work, and engages writer and reader alike in a shared concern for this world. We should say that this style of writing is not immediately or easily adoptable by authors or readers. Certainly, a more personal and subjective way of writing also throws up challenges, not least of perception. How will contributions that value empathetic, impressionistic, experiential and perceptual accounts of living systems be regarded back in the established space of design practice and in circles of design scholarship? We see the personal approach involving words, poetry, image and atmosphere, and often a blending of them all, as a powerful riposte to claims of objectivity and control in design studies. We think that it is one way in which design, and its conventions and foundations, may embark on a path of greater attunement with the natural world. To our minds, embodied and bodily experience, experiential learning, action learning, learning by doing and engaging are all routes that bring us closer to nature and in that sense also to ourselves. That we write about them in subjective terms and the first person is fitting. It reveals that content is congruent with method, that nature is explored through interdependence, mutualism and relationality.

Challenges of working in this area

The work of design and nature, and the deep transformation implied by caring for the Earth, is challenging. It involves profound change to design practice and theory at pretty much every turn and it typically requires unlearning ingrained practices of thought and action as well as striking out in fresh directions. It also compels those engaging with design and nature to try, where possible, to embrace the experience of not knowing and feelings of bewilderment and struggle that are often stirred up in the process. When we editors were working on the first draft of the book proposal for the publisher, the only way we found that we could proceed with committing ideas about this book to paper was to write with our eyes closed. We had to shut out, quite literally, the letters on the page and quieten the thinking mind and instead 'feel our way', more in our bodies, to a different sense of things. Many of the book's contributors took part in writing groups as they were drafting their texts to support their writing process. And many, like us, developed strategies to attune themselves to working inside nature and outside the practice-as-usual world of design. Many felt unsure in a territory that was different from what they had trained in or experienced before. Others took on their feelings of discomfort and found agency and freedom there. When in dialogue with the contributors and the material, we did not talk in terms of contributions to knowledge. Instead we have thought in terms of 'gifts' to an emerging field of design and nature. The gifts are distinctly rooted in the contributors' research and practice; they also stem from their personal experience and from the speculation and imagination we have very much encouraged. We hope the ensuing, sometimes 'unfinished' quality of the contributions will open up spaces for readers to trace new design and nature pedagogies, research and practices.

REFERENCES

- Boehnert, J. (2018). *Design, Ecology, Politics*. London: Bloomsbury Academic.
- Buddhaghosa, B. (2011). *Path of Purification (Visuddhimagga)*. Kandy, Sri Lanka: Buddhist Publication Society.
- Crutzen, P.J., Steffen, W. & McNeill, J.R. (2007). The Anthropocene: Are Humans Now Overwhelming the Great Forces of Nature? *Ambio*, 36(8), 614–621. https://doi.org/10.1579/00447447(2007)36[614:TAAHNO]2.0.CO;2
- Haraway, D. (2015). Anthropocene, Capitalocene, Plantationocene, Chthulucene: Making Kin. *Environmental Humanities*, 6, 159–165. https://doi.org/10.1215/22011919-3615934
- Haraway, D.J. (2016). *Staying with the Trouble: Making Kin in the Chthulucene* (1st ed.). Durham, NC: Duke University Press Books.
- Jickling, B., Blenkinsop, S., Timmerman, N. & Sitka-Sage, M.D.D. (Eds.). (2018). *Wild Pedagogies: Touchstones for Re-Negotiating Education and the Environment in the Anthropocene* (1st ed.). London: Palgrave Macmillan.
- Klein, N. (2014). *This Changes Everything: Capitalism vs. the Climate*. Toronto: Alfred A. Knopf Canada.
- Plumwood, V. (2009). Nature in the Active Voice. *Australian Humanities Review*, issue 46.
- Puig de la Bellacasa, M. (2017). *Matters of Care: Speculative Ethics in More than Human Worlds* (3rd ed.). Minneapolis, MN: University of Minnesota Press.
- Steffen et al. (2015). Planetary Boundaries: Guiding Human Development on a Changing Planet. *Science*. 10.1126/science.1259855.

Glossary

Action learning
A method or approach that combines thinking and actions in context as a form of learning. This approach encourages a continuous and open learning process.

Agency
The capacity of an individual or thing to act, create intervention or influence independently.

Animism
The attribution of a spiritual essence or soul to all natural things and inanimate objects. Animism perceives all things to be animated and alive with influence and agency.

Backgrounding
Reducing the importance of something (such as, in the context of this book, nature or women) by foregrounding other spheres as having agency and therefore dominance. (Plumwood, 2002).

Biophilia
Originally proposed to mean the "love of life" (Fromm), and now commonly used to describe an affinity with, or love of nature.

Con-template—observe together
Consideration with continued attention over a long period of time; watching carefully to the manner something happens or the way in which something is done in order to learn about it, in companionship. (Wernli, Section III).

Conversation—dancing together
The informal oral exchange of ideas, feelings or thoughts that is comprised of a series of back and forth dialogue, often repeated, moving in sync with each other. (Wernli, Section III).

Dialogical
Participating in dialogue in order to understand differing viewpoints, rather than attempting (as through dialectics) to arrive at a single 'truth'.

Dirty design

The dirtiness that comes with bringing together different ways of knowing, challenging epistemological hierarchies, seeking transdisciplinary and multispecies collaborations, challenging single truths or solutions. Keeping design dirty and life affirming by staying situated and close to the soil.
(Tham, Section III).

Earthbonding

A method that includes nature as a participant in all stages of design prototyping – an immersion in nature through design. Human actions with nature to develop a connection to the earth that gives opportunity to understand the integral relationship between environment, design and ecologic self.
(Camozzi, Section IV).

Ecology

The study of interrelationships between organisms and their environment.

Ecosphere

The planetary ecosystem with interactions between all living organisms, non-living components and their environments.

Embodied

Having a physical form, rather than digital or mental; to give an idea or feeling an expression, tangible or visible form. Embodied knowledge is a form of knowledge that is learned through the body's knowledge of how to act, instinctively.

Experiential learning

A process of learning through doing and reflecting, through experiences.
A method of educating through experience, first hand.

Entanglement

A complex and difficult relationship or situation; the action of being entangled, either physically or metaphorically.

Hegemony

The dominance one group has over another that influences and legitimises beliefs, perceptions, ideas and values being accepted as social/cultural norms. This dominance may be so pervasive that it is invisible to many.

Hydroponic

The method of growing plants using mineral nutrients in sand, gravel or liquid, without soil.

Mastery

(a) human control and superiority, or (b) a comprehensive knowledge or skill.

Modernity

An historical period that arose after the Renaissance. A philosophy entwined with the belief that only measurable phenomenon are valuable. Associated with the rise of secular thought, Scientism and Capitalism.

Nature

The 'self-originating material/spiritual world, of which we are a part, including the powers that sustain and govern it'. (Bonnett, 2002).

Organicism

An ancient philosophical view of matter and spirit as united; all beings (including humans) as interdependent without hierarchy. (Merchant, 1990).

Plastisphere

The ecosystems that have developed and evolved to live in human-made plastic environments due to human's overconsumption and waste of plastic products. (Zettler et al., 2013).

Rational anthropocentrism

Interpreting the world through human values and experiences of human beings; human-centred. Argues that humans are the only species able to reason.

Scientific technocracy

The tendency to view science or ecosystems as machine-like systems that could be controlled and managed by humans.

Scientism

An excessive belief in the power of science to solve all problems, or the attempted application of scientific methods, such as itemising and quantifying, to problems that do not call for these methods.

Speciesism

Prejudice or discrimination based on the assumption of human superiority; leading to exploitation of other species.

Social imaginary

The values and symbols set to a particular group that creates their way of living together; represents their social whole and collective life.

Sympoiesis

Becoming with; collective creation, production or organisation; collectively creating and producing.

Vulnerability

The state of feeling open, exposed and unprotected (physically or emotionally). Often described as an emotion.

Wild or *Wildness*

The process that surrounds us all, self-organising nature. (Snyder, 2010).

Wilderness

A specific place, there for the local critters who can inhabit it. (Snyder, 2010).

Word-painting

An action or process of using words as a depiction creating vivid graphic visual work/image. (Landow, 1985; Flaxman, 1998).

REFERENCES

— Bonnett, M. (2002). Education for Sustainability as a Frame of Mind. *Environmental Education Research, 12*, 265–276. https://doi.org/10.1080/13504620600942683

— Flaxman, R.L. (1998). *Victorian Word-painting and Narrative: Toward the Blending of Genres*. Ann Arbor, MI: UMI Research Press.

— Fromm, E. (1986). *For the Love of Life*. New York: The Free Press.

— Landow, G.P. (1985). *Ruskin*. Oxford: Oxford University Press.

— Merchant, C. (1990). *The Death of Nature: Women, Ecology, and the Scientific Revolution* (2nd ed.). New York: HarperOne.

— Plumwood, V. (2002). *Environmental Culture: The Ecological Crisis of Reason* (1st ed.). London; New York: Routledge.

— Snyder, G. (2010). *The Practice of the Wild: With a New Preface by the Author*. Berkeley, CA: Counterpoint.

— Zettler, E., Mincer, T. & Amaral-Zettler, L. (2013). Life in the Plastisphere: Microbial Communities on Plastic Marine Debris. *Environmental Science & Technology*, 47(13), 7137–7146.

SECTION I: LYING

Lying down to receive

The Buddha recommended that we change our body posture in order to begin to see differently. He offered four postures for consideration: lying, sitting, standing and walking. Of these, lying down allows us to be our most vulnerable physically, philosophically and emotionally. This posture is one of letting go of authority and preconceptions. Lying down, we open to receive the world. This is a posture without power and without hierarchies; we can rediscover ourselves as humans among other living creatures.

The state of repose or lying is not one of sleep. Nor is it about passivity. It allows for a quietened state for the mind, one where perceptions can more easily shift. It invites new sensations. In the lying down posture, we can literally have more of our body in contact with the Earth than any other posture. We can contact earthiness as other beings do, and perhaps come to know ourselves as knitted to this Earth.

Lying opens the book with five contributions that invite us to decentre the human and seek other ways to be in the world. The first chapter locates us in the wilds of British Columbia as Louise St. Pierre heeds indigenous advice and explores Buddhist practices as ways of reaching into a relationship with nature as full of creative agency. Louise demystifies animism, the knowing that we can communicate and connect with other beings like trees, ground squirrels and mountains. She interweaves ritual with design practices on her journeys; coming into relationship with the voice and agency of other than humans.

Following this, Katherine Pogson immerses the reader in a poetic voyage into otherworldly days and nights of compassion and gentle engagement with moths in urban London. From her perspective of intimacy with a fragile world, she wonders what it is to be human in an age of ecological destruction. She asks, how do practices that attune ourselves with local species redirect our focus as designers?

In the midst of a Vancouver winter, Avery Shaw narrates how she imagined that she could take charge of mushrooms in her research. Avery found instead that this species took charge of her, changing her daily habits, relationships and the research itself. Mushrooms steer this narration of interdependence and relationship, a narrative that invites us to embrace the abject, non-romantic and less appealing side of nature.

Marcus Dénommé finds surprising places to meditate with nature in the heart of Vancouver. He climbs trees, sometimes to a height of 65 feet. There, he contemplates the contrast between two views of himself. One view is delimited by urgent city activity. The other is an expanded sense of self and relationship he finds when enclosed and supported high in the branches of a tree, and knowing the connection to the root network under the sidewalks. Marcus maps his relationships with the trees against an urban backdrop.

Closing this section, Eugenia Bertulis shares stories of a post-human-centred residency in Italy's Alps. Full of discomfort, questioning and wry humour, Eugenia applies the familiar tools of human-centred design research to honeybees, a deerfly, yeast and a tick. What began as a tongue-in-cheek practice reveals the gravitational force of human needs in mainstream design. Eugenia touches on microbiopolitics, the embodied, visceral lessons of interdependence and the importance of humbly listening to the wisdom of other beings.

Indeed the unfolding of the seasons and our experiences and friendships in nature have unfolded and befriended everything you now see.

Louise St. Pierre

A Shift of Attention

The first time I spoke with a tree was at the urging of a First Nations Elder who simply entreated me to talk to a tree; no further instructions. Hmmm… I had no idea how to do this. So, I started with what I knew: to meditate. I sat in a close old-growth forest and practised slowly returning to my breath until I was ready to speak to the giant red cedar before me. I confessed that I didn't know what I was doing, but that I wanted to talk with her. Almost immediately, I experienced a soft sensation of knowledge that came from both outside and inside of me. I heard "I am here". There was no definable origin for the message. The statement was simple. It was also the truth.

I left that forest profoundly moved and deeply certain that trees can be heard. It was an experience that changed me. I have since taken up practices that allow me to listen, to shift my attention, to be a student of nature. I draw on Buddhist meditation and philosophies along with creative design practices like drawing and journaling. This is how I search for what Buddhists would call the *right relationship* between design and the earth.

> *This is a quiet mountainside. Tender young lodgepole pines crowd softly together. The morning sun backlights their needles, jewels against the mountain across the valley. Slender stems reach eager for the sky. I close my eyes to "hear with my whole body" (Loori 2007). The pulsating forest chirps tweeters calls… hums, bubbles, burbles… everything rustles with gentle rhythms punctuated by the lilting song of the Swainson's Thrush. I hear the Earth herself. She is alive. She is creating.*
> (Journal, May 2018)

Journaling, drawing and ancient practices like Buddhist meditation and philosophical study offer me a path to reflective, non-self-centred engagement with the world. Through these practices, I have come around inevitably to views that can only be described as animist: the certainty that all life forms have knowledge and agency and that humans are interdependent with all these others. I have been surprised at how accessible these animist ways of knowing are.

My practices are varied and can be chosen in the moment. I often begin by drawing trees. I move my hand across the page as I sit with the forest around me. Breathing slows, and the lines just happen: fluid sensuous ink roams across the page. The drawings are not much to speak of, but the practice of doing them is important as it helps to focus attention on the place, the movement, the moment. It is the same with journaling. These journals are a free flow; a bridge between observation, meditation, reflection and study. They are ways of staying present in my body, of paying attention to the particular, of being a student again. I learn this way.

My practice of drawing and journaling is rooted in Buddhist teachings. For many years, I have studied and practiced with a Buddhist community (a Sangha). Here also, I am a student. I am drawn to many tenets of Buddhism: the call for a simple life without resource intensity; the commitment to interdependence as both a philosophy and a lived experience. Vietnamese Zen master Thich Nhat Hanh uses the present tense, we '*inter-are*' (2013), to inscribe interdependence as an ongoing way of being, an ontology, a path, rather than a fixed understanding or theory. Many Buddhist texts declare intentions to support a diversity of beings "whether born from eggs, from the womb, from moisture, or spontaneously" (Nhat Hanh, 1992). Buddhist study helps me see myself as a being among many other beings. It has prepared me for the animist and interbeing practice of talking with trees.

In many indigenous cultures, all entities have been assumed to have their own voice. Trees, rocks and mountains may have a slower pulse than humans and thus various modes of communication (Abram, 2017). But they do speak. As Nhat Hanh (2013) says, "A pine tree is not just matter; it possesses a sense of knowing". The animist belief in the voice and agency of all beings is less mysterious than many assume. Early colonisers could not understand indigenous animism; interpreting animist practices through the lens of scientific rationalism, they assumed that what they saw were primitive beliefs that humanoid spirits and demons could "inhabit and animate material objects" (Plumwood, 2009[2013]: 449). Tales of spirits and possession have distorted the Modern West's understanding of animism. In what might be an attempt to dispel this misconception, Abram (2013) suggests a pragmatic view, that animism may have been a practice that helped indigenous people stay attentive to the unseen qualities and forces of the natural world around them, a world that many of them relied on for survival. Regardless of the context, animism honours the unique agency of other beings, beings that have their own vitality that is distinct from our humanness, and distinct from each other. In that animism respects an otherness beyond our human realm, it is also a spiritual practice.

Like all skills, animist communication has taken me time and practice to develop, but the practice is rather straightforward. It is not so different from how designers learn to read visual language, or how musicians work to become sensitive to tone and pitch. It requires becoming aware of, and then quieting my thinking self, so that I can attune to the differing speeds and modes of communication in the other than human world. Setting aside discursive mind and turning back to the moment, turning attention to the felt sensations, to the body and breath, time after time: this is the practice.

In Buddhist philosophy, we understand this type of direct experience as a greater wisdom, over and above an intellectual or anecdotal appreciation. Mindfulness practice returns me again and again to a sensorial, corporeal, somatic mode of being in the world; a phenomenal ontology that deepens my relationship with nature. At times I pause to touch a tree, or lean the length of my body against hers. Often this brings a sensation of energy or warmth. Sometimes communication. But it is not predictable. My Buddhist practice teaches me this happens, that the communication is impermanent and not to be pursued as an end in itself.

I once met a cedar on the edge of Stanley Park overlooking the sulphur piles on the banks of the North shore. I noticed she was covered in ivy, an invasive species here on the West Coast. I began to peel the ivy away. After a few false starts, it separated from the trunk in glamorous billowing sheets. I tramped the circumference of the tree, pulling at remaining stubs. Suddenly I felt a strange sensation of warmth, a sort of glow emanating from the tree (St. Pierre, 2017). I stopped, unsure of what was happening. It was palpable. Was it an emanation of joy? Gratitude? I am not sure if those would be my humanist interpretations; emotions read on my own terms. And why would I try to interpret this anyway? The tree has a way of knowing and being that is foreign to what Western traditions have taught me.

My practice takes different forms. Some years ago, at a silent lake surrounded by mountains, I tried something learned from Robin Wall Kimmerer (2015): to relate to all beings as named and gendered. I focused on my surroundings, and then as a small incantation, I murmured "Mountain, she is solid on the other side of Lake. Lake, he is as still as glass. Cloud, there you are, a light puff! Loon, she is quiet today".

Suddenly, MountainLakeCloudLoon became powerfully present as a physical sensation. I was overtaken. A full body awareness of another

much larger presence expanded under and through my skin and inside my bones. It took my breath away. I shook myself quickly from that overwhelming moment. I have since understood that this hasty retreat was not a failing on my part: moments of awe cannot usually be sustained (Blenkinsop, 2005). Yet I know a great deal from them.

> *Feeling parched and overheated from the sun on this mountainside, I find a place to do Touching the Earth Prostrations (Nhat Hanh, 2013). Forehead and knees to the hard, dry ground, stinging from tiny stones and pine needles. Ouch. Up. Bow. Down on knees again. Palms to the sky in this ancient Buddhist ritual, I affirm my gratitude for the earth. The Columbia Ground Squirrel nearby is beeping his alarm, a furious pulsating beep. Beep. I stop and notice: Oh! There is a hole right beside my foot… maybe it is his burrow. Stepping away, I take a breath.*

Yes, he is quiet now. Standing and breathing, the world suddenly shifts. I disappear. I am air. I am a breathing part of this forest. My *chest expands and I float with the air that slips around me, this hot earth, this pine tree, that ground squirrel.*
(Journal, Manning park, July 2018)

In that moment I felt both interconnected (at one with) and charged by the unknowable outside of myself (expanded). It was what Abram (2017) describes as "an alterity that exceeded my own sentience... a connection with something unknowable, an enigma that drew me into earthly relationship". I became aware that there was no separation between me, earth, trees and ground squirrel. I understand that many First Nations communities do not consider our human skin as a boundary or separation between humans and other beings (Beeman & Blenkinsop, 2008). But even as there was no separation, I did not lose myself; interconnection was neither a unification or a dispersion, but a humbling and heart-opening sense of myself among others. It was what Bai, who researches contemplative pedagogy, calls sympathetic resonance, "the perceiver's participation in the perceived… a communion, a transfusion between them" (2001: 21).

In Buddhist teachings, relationships become a third knowing: "You and I are the same thing, yet I am not you and you are not me… There is a reality that transcends all dualities" (Loori, 2007: 36). I felt opened to a space outside of myself, a knowing beyond being human, a great wonder. I was in relation with beings I perceive as completely different from myself. It was what Abram (2017) calls "a genuine magic". Children who have not been conditioned into Western dualities can access this magic more easily than adults. One nine-year-old has described her animist conversations exactly like mine: "Little words curl into your mind" (Blenkinsop & Piersol, 2013). Astonishingly, she went on to specify: "You have to know that you're not thinking" (ibid.).

Buddhist practices help me identify when I am thinking or not. My design practices of drawing and reflective journaling also take me from my thinking self to my body, and help me let go into a wonder of animist awareness. Many design practices can support other ways of knowing. Sketch models, drawings, reflective documentation, role play and storytelling can unlock designers from their thinking self and help them see things differently. Like animist practices, they shift the body and "spread mind and creativity out much more widely" (Plumwood, 2009[2013]). In design, we use these methods to seek insights and to explore new relationships. As designers we can choose what we pay attention to. We can set our intentions beyond the human-centred. These design practices are only a slight remove, requiring a mere shift of attention to enable animist learning.

It is a small turn, a shift of attention maybe just 20° away from what we have been taught to seek, to bring attention to the living earth; to forests, ravens and ground squirrels. To bring ourselves to something larger than ourselves, to see that everything we do, in design or otherwise, belongs to the earth. This is *right relationship*. Animist awareness brings a fierce questioning of Modern Western priorities in design. It closes the door on our old priorities of commodification, individualism, acceleration and distraction: once we know the reality of the natural world, there is no going back. Animism has allowed me a route into a fuller understanding of my place within and dependence upon nature; of *our* shared place within and dependence upon nature.

REFERENCES

— Abram, D. (2013). The Invisibles: Toward a Phenomenology of Spirits. In G. Harvey (Ed.), *The Handbook of Contemporary Animism* (pp. 124–134). Durham, UK: Acumen Publishing Ltd.

— Abram, D. (2017). Earth and Dark Wonder: Notes on Animism and Technology in an Age of Ecological Wipe-Out. Public Lecture, Department of Curriculum & Pedagogy, UBC, Vancouver, July 19.

— Bai, H. (2001). Challenge for education: Learning to value the world intrinsically. *Encounter*, 14 (1), 4–16.

— Beeman, C. & Blenkinsop, S. (2008). Dwelling Telling: Literalness and Ontology. *Paideusis*, 17(1), 13–24.

— Blenkinsop, S. & Piersol, L. (2013). Listening to the Literal: Orientations Towards How Nature Communicates. *Phenomenology & Practice*, 7(2), 41–60.

— Blenkinsop, S. (2005). Martin Buber: Educating for Relationship. *Ethics, Place & Environment*, 8(3), 285–307.

— Kimmerer, R.W. (2015). *Braiding Sweetgrass: Indigenous Wisdom, Scientific Knowledge and the Teachings of Plants*. Minneapolis, MN: Milkweed Editions.

— Loori, J.D. (2007). *Teachings of the Earth: Zen and the Environment* (1st Shambhala ed.). Boston; New York: Shambhala; Distributed in the USA by Random House.

— Nhat Hanh, T. (1992). *The Diamond That Cuts Through Illusion*. Berkeley, CA: Parallax Press.

— Nhat Hanh, T. (2013). *Love Letter to the Earth*. Berkeley, CA: Parallax Press.

— Plumwood, V. (2009[2013]). Nature in the Active Voice. In: G. Harvey (Ed.), *The Handbook of Contemporary Animism* (pp. 441–453). Durham, UK: Acumen Publishing Ltd.

— St. Pierre, L. (2017). Bioinclusive Pedagogy: Redo Our Relationship with Nature. *Cumulus Kolding Proceedings*. Retrieved from http://cumuluskolding2017.org/proceedings/

Katherine Pogson

'Towards ... Something More Liveable'[1] *A Moth Journey*

It begins with a place. A chalk-hillside, June. While my partner photographs rare native orchids, I start to observe the insects more closely.

It begins with a book. Thousands of colour photographs of living moths (Manley, 2008). Startling, disregarded, numerous—yet in steep decline. Gateway to another world, I still have not tired of this book.

It begins in the studio. A project responding to vessels in The Pitt Rivers Museum archive begins to 'go all moth'. The textiles connection is obvious, reflected in the common names: Cloaked Carpet, Netted Pug. The structures begin to seep into my practice—yet I know it is not really about form. Colleagues and museum staff alike, react with bemusement, "Don't they just eat your clothes?" I wonder what I am attempting to work out through this process.

And so, it begins. I begin to take moths personally. I stand up for, and with, moths.

9pm Hornsey Rise:

Up the ladder to the roof.

The metal frame stretches up from my first-floor terrace onto a flat roof. I climb over the ledge, past the satellite dish, pulling my bag of cables, wooden slats and plastic jars behind me.

The North London evening sky is overcast, with the orange glow of sodium lights and intermittent glimpses of moon.

In the act of stepping out onto this platform, I enter a different world. I can see into kitchens and living spaces, crowded together, remnants of a garden, the back yard of the pizza shop.

Lime-speck pug. North London roof. 12 May 2017.

Immediately I feel the wind, the elevation of the hill and sense the cardinal points—sunset and sunrise—in a way that I am woefully unable to do at ground level.

Unreeling the electric cable, I cast it off, lowering it through a skylight to a plug in the bedroom below.

A simple plywood box, with two angled sheets of Perspex over the top. The cable attaches to a fluorescent actinic bulb, which gives off an ultra violet glow and should not be looked at directly in case of retinal damage.

Feeling slightly ludicrous in my night-time sunglasses, I wait for the first wave of insects, the dusk flyers.

And here they come.

Their eyes shine copper in the torch light. You feel rather than see their approach, in a whirring set of wing beats, a percussive sensation on the ears.

Fragility is not the impression you receive, surrounded by dense, circling bodies, as the evening deepens. It is more a sense of urgency, of force of intent—as the moths home in, repeatedly diving towards the light.

6.15*am*   *Hornsey Rise:*

The struts of the ladder are sweaty with dew. The sun is already high enough to warm one side of the box.

And now the urgency is on my side. A cascade of diverse creatures, mostly inert, cling to the sides of the structure. A pale green geometer, wing-tips touched with crimson, escapes before I can photograph it. A huge Hawkmoth, the shape of a stealth-bomber, allows me to lift her on a piece of paper. She rolls back her grey wings to reveal bright blue eye markings on a flash of pink.

Panicking slightly as the heat rises, I feel a growing sense of responsibility. The roof is devoid of vegetation, exposed, and the vulnerability of these creatures in daylight is clear.

Birds gather, knowingly. An apparently expired moth lies on its back at the bottom of the box. I want to record, shelter and free them as quickly as possible. This short intervention of a few hours begins to feel like a trespass.

6.50*am*   *Woodberry Wetlands:*

At the nature reserve nearby, I have started a moth recording group. We monitor population levels and distribution, as many species plummet in number. A familiar story. Light pollution, pesticide use and climate change all play their part in a complex picture, but simple loss of habitat is key.

Cycling there through Finsbury Park in the early morning, I see my local patch from a different point of view. Noticing tents among the shrubbery, night workers asleep on the benches, I gain an impression of my neighbourhood, occupied in waves of time by different shifts of humans.

The reserve is a reed-fringed reservoir surrounded by high-rise flats. Rounding the corner, the sound of birdsong is instant. The sun blazes through the tops of the reeds. The moth population here is quite different from my home a mile and a half away, and I begin to learn more about the intimate, evolved relationship between geology, water, specific food-plants and particular species. The seasons pulse with native and migrant birds, insects and humans, synchronised to the rhythms of vegetation and weather.

8.30am ***Blackstock Road:***

Down Fast Fashion Alley on the way to work, Arabic sequined gowns swing and Romanian lorries unload. In search of breakfast, I pass a series of food outlets: Lebanese, Uighur, Turkish, Ethiopian, Japanese. Within the almost instant access to global resources of an urban setting, food seems to be the most essential reminder of home.

Musing on my locale in terms of sustenance and supplies, I develop the thought that these synthetic imported textiles might be my most 'local' materials.

It occurs to me that the moths—those highly specialised, invisible creatures often thought of as a devouring plague—are simply being starved out of existence. What if they are famished?

9.45am ***Studio:***

> *Conversations among ourselves have always had other participants.*
> (Ghosh, 2016)

The objects I make are refusing to be accessories any more. They no longer want to talk to or about the human body so directly. I produce a series of very wrong things as this dialogue plays out.

Relinquishing habitual outcomes, materials and processes, I experiment with textiles in different modes and scales. My new subject matter begins to unfold thematically in terms of nourishment, procreation, shelter—appropriating the Maslow[2] pyramid basic hierarchy of needs for nonhuman ends.

As my focus shifts, I look for collaborators and other outlets—writing, installation, workshops—ways of communicating through an expanded practice, the journey in itself.

Decentring the human has liberated me from the artefact, too in a way.

Conclusion

> *wayfinding, then, more closely resembles storytelling than map-using*
> (Ingold, 2000)

The narrative of my 'moth journey' leads from seeing nature as a resource, through attraction and curiosity, to direct observation, sensory enhancement and learning. This fosters a growing sense of empathy and ecological responsibility, which in turn prompts renewed cultural engagement and material action.

But what does it mean to 'stand up for, and with, moths' as a mode of creative practice?

If the problem of the Anthropocene is one of how to be, rather than how to act (Maggs & Robinson, 2016) examining the purpose of design in an age of ecological destruction requires a profound rethinking of what it means to be human itself. One role for design might be to develop practices which question human boundary constructs.

Survival, or "ongoingness" (Haraway, 2016), entails fostering a sense of 'self' which includes intimate 'kinship' with nonhuman nature. Inhabiting this more porous sense of being alive, we may begin to experience our actions as a form of 'self-harm'. Practices such as these may help to dissolve the cognitive rift which seems to paralyse privileged nations from *feeling* sufficiently the negative effects of human actions, which we so efficiently document (Conrad et al., 2006).

This journey embodies two paths: one which leads away from making, and one that returns to it.

The first path requires "*disciplined thinking combined with strategically, profoundly, madly letting go*" (Meadows, 1999). Intriguingly, this implies liberation rather than denial—the forward momentum of moving on from practices which societies have outgrown.

Part of this will entail relinquishing materials and habits which are limiting and harmful. More ecological ways of relating trigger changes in consumption behaviour, political and cultural expression, which are design outputs in themselves, with or without artefacts.

For design, this suggests an uncoupling of the link between established economies of desire and new ideas about value, to redirect the creative urge outward and away from production, towards community engagement, resilience-building and knowledge-sharing.

The second path requires a deeper engagement with materiality.

Design solutions reliant on industrial processes create 'simplified ecologies'—a symptom of "life-world disengagement" (Tsing, 2016). Accepting the contradictory, uneven 'muddle' of the present, paradoxically releases blockage. For complexity and entanglement suggest richness, biodiversity—hope.

Practice implies habit. My exercises in sensitivity, or "attunement" (Morton, 2018) begin, simply, with what is close by. Engagement with my local 'place', its inhabitants and processes, has given me an entry point into a joyous world of layered themes, images, activities and connections.

Aware of the anthropomorphising potential in imagining 'the moth's point of view' I nevertheless find value in this starting point.

I appropriate the textiles vocabulary of repair, care and human domesticity, to speak about a creature usually associated with damage, nuisance and contamination. This inversion allows me to explore overlooked stories of neglect, destruction and unintended consequences.

Through this approach I attempt to untangle and redirect what is materialised through the act of making in order to ask the question 'what truly nourishes you?'

NOTES

1 Donna Haraway at AURA conference: "Anthropocene: Arts of Living on a Damaged Planet", Santa Cruz, California, 8–10 May 2014. Anthropocene, Capitalocene, Chthulucene. Available at: http://vimeo.com/97663518 (accessed 18 January 2016).

2 Maslow, A.H. (1943). A Theory of Human Motivation. Psychological Review, 50(4), p. 370.

REFERENCES

- Conrad, K.F., Warren, M.S., Fox, R., Parsons, M.S. & Woiwod, I.P. (2006). Rapid Declines of Common, Widespread British Moths Provide Evidence of an Insect Biodiversity Crisis. *Biological Conservation, 132*(3), 279–291. doi: 10.1016/j.biocon.2006.04.020
- Ghosh, A. (2016). *The Great Derangement: Climate Change and the Unthinkable.* Chicago, IL: University of Chicago Press.
- Haraway, D.J. (2016). *Staying with the Trouble: Making Kin in the Chthulucene.* Durham, NC: Duke University Press.
- Ingold, T. (2000). *The Perception of the Environment: Essays on Livelihood, Dwelling & Skill.* London: Routledge.
- Maggs, D. & Robinson, J. (2016). Recalibrating the Anthropocene: Sustainability in an Imaginary World. *Environmental Philosophy, 13*(2), 175–194. doi:10.5840/envirophil201611740
- Manley, C. (2008). *British Moths and Butterflies: A Photographic Guide.* London: A & C Black.
- Meadows, H.D. (1999). *Leverage Points: Places to Intervene in a System.* Hartland, VT: The Sustainability Institute.
- Morton, T. (2018). Being Ecological. London, RSA event, 29 January 2018. Available at: https://www.thersa.org/events/2018/1/being-ecological
- Tsing, A. (2016). Earth Stalked by Man. *The Cambridge Journal of Anthropology,* 34(1), 2–16.

Avery Shaw

An Unexpected Rapport: Mushrooms, a Designer + Everyone Else

A few years ago, I quit my job as a product designer for the (human) home environment to return to school to focus on sustainable design. I started growing mushrooms as a research study of material exploration and human–object relationships. The proposed project was to grow mushrooms on everyday household objects to give to participants. My hope was to gain insight into the connections people have with objects when the notion of care is attached. In order to create these "mushroom objects" for participants, I had taken time to experiment and focus on learning the capabilities of mycelium and mushroom growth by manipulating them. However, by the end, it was me being manipulated. Whatever the influence was for this shift (and there were several), I was changing how I approached my design practice. While the growing process was an experience, it was more the actions and themes happening around the mushrooms that would be the most relevant.

The process became so much more than just growing mushrooms as a material but turned into a relationship, a rapport between the mushrooms, myself and everyone else.

And isn't that a defining aspect in design—the everyone else?

Everyone else—the narrative

The everyone else—the aspect of influence, the aspect of narration. How do I form stories for others to create meaning in my actions? I was naive to the power of the narrative. How it influences, how it directs and guides. How it can be manipulated, transformed and moulded to tell a certain story (Bertolotti et al., 2016). The beginning of my mushroom narrative was neither. I wasn't aware of the narrative, that there was a narrative, until it would not be ignored.

The unacknowledged narrative started with purchasing. Homestead Junction was a shop on East Hastings[1] in Vancouver that sells mycelium blocks from a local supplier. They are the only store in Vancouver to sell mycelium blocks.[2] However, their storefront had closed just days before I went on my purchasing expedition. A notice on their door stated they were restructuring their business to focus on workshops and their online store. When you ordered online, your purchase was available for pickup from their neighbouring store — a record shop.

This was the moment when the narrative refused to be on the sidelines, to be ignored, or be the second focus. From this point, as I started explaining my actions to others, I noticed the conversations around the mushroom. Starting with my mother, she had some concerns. The more I tried to explain the situation the worse it sounded — I was picking up mycelium blocks to grow mushrooms from a store on East Hastings that had closed, but I had arranged to pick them up from a record store next door. Don't ignore the conversation — the narrative. It is how others perceive your work, understand it.

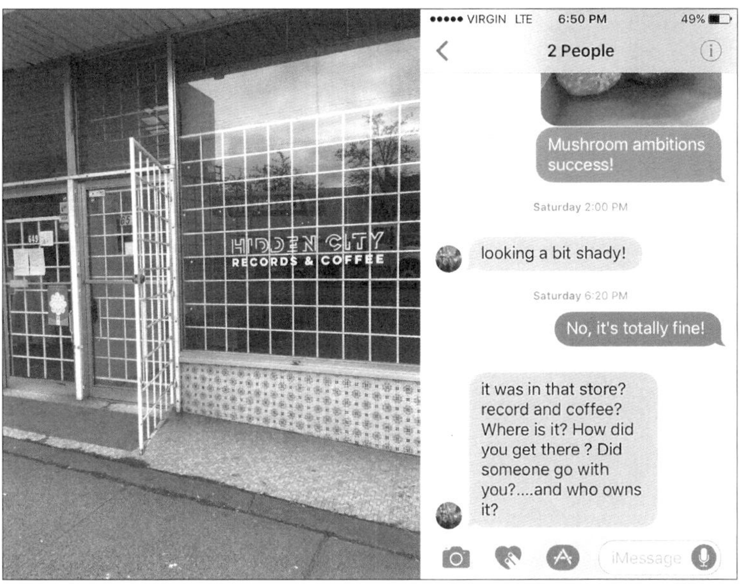

My mum started to refer to this as my "mushroom ambitions". My farm-raised father was giving me advice to growing them on manure piles. My friends were calling me a hippie, that "you're so west coast now". And the "shroom" comments, there were a lot of "shroom" comments. From my international cohort peers, most surprisingly, I was receiving comments about their experience with them—how they'd been foraging for them or knew people who grew them. I came to realise that in North American culture, mushrooms can have a certain set of assumptions attached to them (like "shrooms" or immediate association to the dirty, smelly, mouldy), that I have come to assume is from our limited interactions with them. The conversation that surrounded my 'mushroom endeavours' highlighted for me the assumptions, preconceived notions, and expectations from a cultural/social influence. Not just specifically towards the mushroom medium—but on a wider scope the assumptions, preconceived notions and expectations we, as people, bring to everything.

Sharing (personal) space

Which led to the influence cultivation took on my personal space—the assumptions, preconceived notions and expectations I brought into this shared space between the mushrooms and I. Mushrooms are not animals or similar to plants, which I naively assumed they were. They are their own kingdom. I purchased the first mushroom block in February (the middle of winter) when growing them outside was not an option. Mushrooms grow best in colder temperature between 10–15 degrees Celsius and with 10 hours of indirect sunlight—ideally a north-facing window. It became key to their survival to find the perfect spot for the mushrooms to grow. In my studio, I found a space that met their ideal growing conditions—a classroom. At home, I adapted my living space to their needs. I turned the temperature down in my studio apartment and opened the window to get the temperature to 15 degrees

Celsius. Adapting myself and space to their needs. I had to wear multiple layers at home — two bulky knit sweaters, two pairs of socks and a toque.[3]

Living in close contact with this "other" organism challenged my sense of control over my built environment — that home environment that I had spent years designing products for.

Some blocks were slow to grow, while others developed overnight. Some oozed smelly "mushroom juice", while others would dry out. They were unpredictable in how they grew. I tried to anticipate and understand their needs. In their adapted space (my home), I became invested in the growth process, while the ones in the found space (my studio) were neglected. I adapted to the lower temperature at home and did not change it back even after the mushrooms had grown (and "moved out"). I adapted. I found the colder temperature more comfortable after weeks of mushroom growth. This interaction ultimately changed my own habits to accommodate them. The act of growing takes on a level of care that influences the relationship. Anthropological archaeologist Uzma Rizvi (2017) reflects on the influence of care stating, "ultimately, that care for and with others is also self-care. Once we recognize ourselves, we begin to recognize our positions, and how our positions may be at the expense of others, be those others human or non-human". When the notion of care becomes involved, we shift our approach. When that notion of care is tied to our environment, we adapt our approach (and ultimately our perspective).

Living with mould, rot and decay

Then there was the mould, rot and decay. Mushroom cultivation is not for the weak stomached. Mould is common, rot and decay are inevitable. One has to contend with them, as they come hand in hand. Again, I felt like I was being manipulated by the mushroom medium — no longer in control of the outcome. While care, control and expectation were implicit in growth, they seem to be lost when mould, rot and decay took over. At first, I was anxious and hesitant to touch, to interact, with the blocks/objects as they began to disintegrate — they felt fragile. That also came with a sense of sorrow, denial. I closely witnessed the mushrooms, that I had lovingly cared for and developed a relationship with, mould, rot and decay. Out of my control, as they started to crumble and deteriorate. The mushrooms smelled, they gave off mushroom juice, brown sticky liquid. That I soon became numb to, adapting to. No longer hesitating to touch.

I was approaching the mushroom as a material, and I had never worked with a material with such a short life-cycle (in the human sense of time). I was forced to deal with the lifespan of materials directly. I had to change my view of the mushrooms—they were not the stable, reliable, steadfast material I was used to working with. I could not approach it as a material.

Adapting—growth

So, what can we learn as designers from actively engaging and consciously existing with the natural world? What can we learn by paying closer attention through observing our social-constructs with the natural environment?

My approach as a designer completely shifted when I began to physically encompass elements from the natural environment into my work and research. When I began working with and caring with the natural environment, I began to consider our environments and the emphasis between our spaces became evident: the separation of our built and natural environment, and the mixture that does, in fact, exist. The mixture of the built and the natural, human and environment, indoor and outdoor, the smooth and the striated (Deleuze & Guattari, 1987). Through this work I noticed that when we come into our built environments, we are not leaving the natural one behind—that is the illusion of control, the illusion of us & them. The social-construct that has heavily influenced our society.

Was I able to get mushrooms to attach to the mundane object? No, not in a consistent manner. I had to adapt as a designer. Developing my practice, shifting my process, growing as a designer—working with the dirty, the nitty-gritty.

As I'm finishing my masters at Emily Carr University of Art + Design, I am still purchasing the mushroom blocks. Homestead Junction is now fully closed, meaning I purchase the blocks directly from the supplier. A supplier (with the online domain—shroomstore.ca), who will deliver right to your door in a white van accepting only cash. While this sounds dodgy, no one seems to make comments anymore. As if this is what is expected now—adapting to this narration. Adapting to environments. Adapting as designers. Growing.

NOTES

1 East Hasting Street is in one of the poorest neighbourhoods in Canada, known for high levels of drug use, poverty, mental illness (MacQueen, 2014).

2 A 5lb block of compressed sawdust/wood-chips/limestone injected with mycelium, sold in clear plastic bags with a cotton wad on top.

3 A Canadian woollen hat.

REFERENCES

— Bertolotti, E., Damm, H., Piredda, F. & Tassinari, V. (Eds.). (2016). *The Pearl Diver: The Designer as Storyteller* (vol. 1). Milan, Italy: DESIS Network Association —Dipartimento di Design, Politecnico di Milano.

— Deleuze, G. & Guattari, F. (1987). 1440: The Smooth and Striated. In *A Thousand Plateaus: Capitalism and Schizophrenia* (pp. 474–500). Minneapolis, MN: University of Minnesota Press.

— MacQueen, K. (2014). What Rules Apply in Vancouver's Downtown Eastside? *MacLean's*. 24 March. Retrieved from: http://www.macleans.ca/news/canada/what-rules-apply-in-vancouvers-downtown-eastside/

— Rizvi, U.Z. (2017). Decolonization as Care. In A.P. Pais & C.F. Strauss (Eds.), *Slow Reader* (pp. 85–95). Amsterdam, NL: Valiz.

Marcus Dénommé

Sitting in Trees

Back support:

Just a few blocks off my commute home is a small neighbourhood park scattered with trees. They all blanket the ground with red and yellow, leaving the trees bare. I toss my bag and sit down, leaning against the first tree I come upon, a tall White Alder. Immediately, I feel the details of the tree up against my back, it's actually a comfortable seat. A seat that perches me upright, sitting tall with all the back support I could ask for, unlike the stools at work, or the seats on the bus. It's not even 4 pm and I start going over all the different chairs and seats that I've sat in today—like a frantic game of musical chairs.

I didn't come here to ponder the ergonomics of furniture though, I didn't come here to ponder anything. I am a practicing printmaker, passionate about the world of information dissemination, obsessed with text and image and how multiplicity changes the meaning of a message. Despite all this, the part of my practice that feels most worth sharing are the moments where nothing happens at all. I've been venturing away from busy streets, and taking detours to visit trees around the city. Although sometimes brief, these habitual interactions have become an integral part of my daily shuffle. Like today, for example, back support is not the only kind of support that I'm getting from sitting here. Having a moment to slow down and ground myself—literally and metaphorically—alongside a tree is the main reason I'm here.

This tree might as well be a wall. It reminds me that my shoulders can stretch wide, and my chest can open again. Leaning against something so strong and solid gives me a moment of certainty that I won't fall backwards. From here, I can see the widest view possible, listening to the faintest noises. I feel the leaves crunching under me and wonder where a particular smell is coming from. Looking down I see that the local ants have come to investigate my pants, by the dozens.

Pine tree at the base of China Creek North Park.

Room to be:

I don't notice the loud noises of traffic, horns and sirens when I'm surrounded by the tree, as there is a moment of silence. With a moment of silence, I notice the absence of traffic noise, and the overwhelming sensorial stimulation that is always around us. I imagine most people, including myself, build a resilience to the constant noise of the city to keep from jumping with every claxon, or flinching at the passing transports. This acclimatisation to urban environments has affected my ability to connect with nature. Yet when I can let my guard down, I notice how much nature intertwines with urban settings, like this one.

I discovered a tall Cedar tree tucked into the courtyard of an apartment complex. While the first 15 feet of the tree had no branches, the rest was dense and bushy. Hugging the base of the tree is a concrete footpath, which is probably why the tree had been pruned, so as to not obstruct pedestrians on the walkway. Ironically, the roots underground managed to undulate and warp the pathway as though an earthquake had hit. I am amazed by the tree's height, and surprised to find the giant tucked into this part of town. My curiosity leads me up the thick spongy bark, and I crawl into the first branches. Inside, is a dramatically different setting. The cedar needles are thickest at the end of the branches, making the inside spacious and well sheltered. The sun shines through with spotted diffused light. The branches are bouncy and flexible like a trampoline. I can hardly see the ground below me, and all the harsh city noises from outside are pleasantly muffled. The longer I sit here, I can feel my body shedding the layers of resilience built up from the sensorial stimulation of the urban environment. Here, there is room to *be*. Just be.

Cedar tree next to Fountain Way pedestrian path.

I notice that when sitting in a tree, my senses grow and expand. I work differently here, no longer confined by the walls of a workplace, or the sidewalks of our urban grid, the more open space helps me pursue creative thoughts and processes that do not ignore biology and nature, but are directly informed by it. I can learn from this tree. Less so by studying it scientifically but actually interacting with it, feeling its surface. The longer I sit in this tree the greater a relationship I build with it, and all the wildlife that inhabits the tree. This bond with the biology and hydrology of the environment is crucial to my practice and ability to design. I can only imagine how old this tree is, and with that are endless deep rooted teachings connecting to the shores of the Salish Sea.

Hanging report:

Date	*Location*	*Weather*	*Repeated Tree*
Monday February 19, 2018 2:00 pm	31st and Ontario St	2°C Sunny with snow covered ground, below zero in the shade.	Yes

Tree type	White Alder.
What did I leave	A small wooden plaque (3.5" × 4.5") nailed to the tree.
What did I take	Deep breaths.
What did I learn	It's a challenging skill to become quick at swinging a sling around a tree branch to secure in order to climb higher.
What did I unlearn	That the shape of this city still has ridges and valleys carved from the rivers that use to flow through here.
What I saw	5 km north are the foothills of the north shore mountains, soaking in the sun. In all the notches of the branches were a small tuff of snow and ice, that remain from the last snowfall.
What I heard	Distant noises from the city muffled, and the north wind trickling through the trees around me.
What I felt	Excitement, and focus. I felt the distance between myself and the ground.
What did I smell	The smell of dry moss that is rubbed off a tree. It floats in the air like dust and for a moment it engulfs your nostrils with a musty aroma.
What did I remember	Childhood stories.
Height climbed	25 m
Total time in tree	3.5 hours

65 ft:

I rest my weight back into my harness and let go of the branches. One harness, four nylon slings and six carabiners have become my frequently used tools for doing this work. This is the part of my practice that pulls me high off the ground, and into the tops of trees.

I've become obsessed with big old Alders. They resemble a concrete pillar with branches exploding from the trunk. At this time of year, these deciduous trees are bare, exposing every inch of the grey, rock-like bark. There are five Alders in this park, but I approach one in particular that stretches high up above the rest. I loop a nylon sling from the branch above me and clip it to my harness, securing myself. From there, I move up to the next branch and repeat this process, which means I'm always 'anchored' to the tree, and can safely ascend. An hour passes as I make it three quarters of the way up. Throughout the climb, I focus on keeping my breathing steady. The movement is slow and meticulous. I press my body to the tree to balance as I move up each branch. I now anchor myself to two points of contact on the tree, and let go. My feet, arms and legs dangle as I sit back in my harness. Now with my hands free, I get to work (see *Hanging Report* above). Dropping a long roll of cord tells me that I am 50 feet off the ground and so I keep on moving up, climbing higher than I have before. As I make it to the 65 foot mark, I grab the hammer and nails that have been dragging behind me and carefully hammer a small wooden plaque into the tree. The bark so dense that the noise from each strike of the hammer echoes across the park.

When I look straight down to the ground, I feel a steep rush of adrenaline. The longer I sit here, the butterflies in my belly disappear one by one. These slow interactions with trees have a way of bringing up old memories. The last time I was here, I remembered the childhood stories my father told about a catastrophic accident he faced as a 19 year old. He was working on an oil rig in 1982 when he was knocked from a platform and fell 65 feet to the ground. He spent months hospitalised, with permanent injuries to his body. What was most significant to me from those stories was the height at which he fell: 65 feet. As a kid, I would constantly scan buildings and wonder how tall they were in relation to 65 feet, to better understand and visualise what my dad had experienced.

Installation on White Alder tree on the north east side of Queen Elizabeth park.

It's now February 2018, and this memory inspired me to design and install a piece artwork at 65 feet up a tree. A landmark to the geographical location and height at which I placed the plaque. "**65 FT**" is carved in bold type into the wood from a CNC machine. Below, a land acknowledgment to the xʷməθkʷəy̓əm, the Tseil-Waututh, and the Skwxwú7mesh Nation. Like a giant pillar, its roots are strong and tightly intertwined with the rocks and soil below. Without these root systems, the weight of my body sitting in its branches would make the tree topple over. Yet the higher I climb, the more reliant I am on the root systems that spread below our feet.

REFERENCES

— **Vancouver (B.C.). Engineer's Office (1963) *Map 336—The Official Street Name Map of the City of Vancouver* [online]. Vancouver: City of Vancouver. Available at: http://searcharchives.vancouver.ca/official-street-name-map-of-city-of-vancouver [20 March 2018].**

Our email messages back and forth during the book's planning and making are the book's nature almanac. They accompanied our messages about contracts and deadlines. They connected our lives as they do the work of design.

Eugenia Bertulis

Co-creating with a Tick

It was an absurd ride, the Italian mini-bus tearing up the road toward a high valley famous for regionally protected cheeses. My research proposal had been accepted by the Nature Art and Habitat residency in Taleggio, in the Italian Alps north of Bergamo. The trip felt quixotic. Waterfalls and cliffs flickered past the bus windows. I was barely sure I was on the right bus, not completely comfortable that the driver understood where I wanted to get off, and jet lag had thrown off my sense of time and place. My research proposal was speculative, applying the premises of human-centred design research to research with non-humans, and I doubted that it made any sense.

I had written the proposal as a response to the philosophical provocation that non-humans might have agency. Many writers are revisiting questions of non-human personhood. How could such an epistemological shift, from anthropocentrism to more relational conceptions of our world, influence what we do as designers? How could the quasi-object (to use Bruno Latour's (1993) dense construction) help a practicing designer move beyond biomimicry? How could we ensure that when a designer finds inspiration in, say, honeycombs, that the resulting structures aren't built in fields which displace the inspirational honeybees?

The format for my proposed research was simple: I would take familiar design research methods and carry out investigations with non-humans as the participants and co-creators. As Sanders and Stappers describe in "Convivial Toolbox" (2013), methods such as co-creation, participatory design and cooperative design research date back to the 1970s. These methods mark a considerable shift away from earlier exploitative models where the researcher compiles knowledge via passive test subjects. Today, participatory design research involves reciprocal designer/participant relationships. "In a co-design process, the roles change: the person who will eventually be served through the design process is given the position of 'experts of their experience', and plays a large role in knowledge development, idea generation and concept development" (p. 24). I began with these premises; that non-human agents might also be experts of their own experience, and that human design work might serve non-human agents as much as human ones.

Media-release forms, observational research and the importance of listening:

After my mini-bus paused to drop me in Sottochiesa, I found my apartment and reviewed my prepared research materials. My first activity, a bit tongue-in-cheek, was to take the official Canadian Tri-Council Research Ethics materials and ask non-human participants to sign a media release form and an informed consent form. Honeybees seemed the right community to approach; humans have an ancient relationship with honeybees, dating to Roman times in this valley, and we regard them with concern and affection. In the morning, I set off through forests, orchards and pastures to find some participant researchers. A sunny half-day of wandering later, I found a colourful row of beehives behind a shuttered stone cottage.

How to describe the futility of my project at that moment? I sat in the mossy grass at some distance. I watched the beehives, some empty, some with thick swarms curling around their sides. I grazed on wild strawberries, and felt as intimidated as a first-year design student instructed to approach a stranger for an interview. The bees seemed angry in their buzzing. Earlier in the day I had found a snail in the woods, and tried to entice it to research by offering it a leaf as food. As I watched the honeybees, I thought about communication in research; my problems communicating in broken Italian with the bus driver and neighbours. My nourishing leaf offered to the snail reminded me of the bowls of candy and pretzels on the tables at old-style focus group research sessions.

Eventually, I left the honeybees in peace. It seemed important to let my attempted recruitment fail, no matter how fruitful and poetic the interview might have become. I wanted to pay attention to non-verbal cues. I often forget the importance of listening, and have the socially indoctrinated habit of filling silence with explanations of my design intent. When doing research with human co-creators, I have to remember to let confusion expand, awkwardly. I have to practise replying "what do you think that confusing design feature does?" and listen for insights. At that moment, while keeping my distance from the honeybees, I had a chance to practise listening for other forms of wisdom.

A few days later, as I sat in a meadow a few kilometers away conducting another research study (which involved playing a variety of cricket sounds from a downloaded app to the hidden crickets in the field), a deerfly came and landed on my media release form. I took a photo and publish it here.

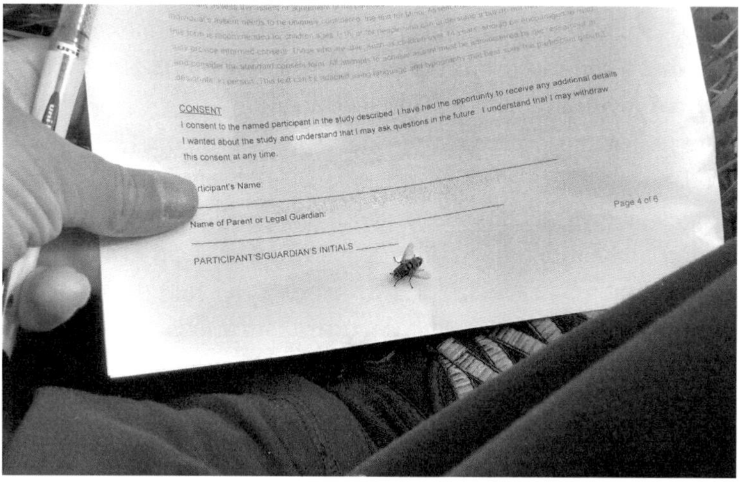

Deerfly and media release form.

Informed consent, co-creation and time:

In legal and animal-rights contexts, the issue of working with those unable to communicate informed consent is sometimes addressed by assigning proxies or guardians-ad-litem. The Whanganui River in New Zealand made headlines in 2017 for winning legal personhood status. Humans now are representatives on its behalf. Invoking these precedents, I asked some fellow design residents in Sottochiesa to collaborate with me on my next research activity. I hoped to get permission from an energetic stream nearby, the Enna, to collaborate on some co-creative activities. They gladly agreed to sit with me at rivers' edge below a waterfall. They were experienced in meditative practices, and found it moving to contemplate and discuss what wisdom the Enna might have been imparting to us. When I asked if they felt the Enna would agree to participate in a co-creation workshop I had prepared, they assented on behalf of the stream. We passed around the informed consent paperwork that I had brought with me from Canada, somewhat self-conscious and laughing at the absurdity. The bureaucratic became ceremonial in its formality. As fellow designer Louise St. Pierre (2015) writes on the importance of bringing insights from Buddhist, First Nations' and animist practices into design work: "Ceremonies invite reflection by offering a moment of pause, again and again. Ritual and ceremony are known as *form* in Buddhist practice, referring to the shape and pacing of moments of ritual, pause, and noticing. These forms provide a container for seeing things anew".

The paper *forms* of *informed* consent gave us designers a meditative *form* and timing for practising design research.

After completing the prerequisite paperwork, we continued with the activities I had prepared. One activity involved filling small glass jars with water, sand, moss, and duff to bring back to Sottochiesa. There we opened the jars with neighbours to smell and feel the qualities of the river, sketching and designing for multiple senses and diverse forms of knowledge and wisdom. The idea of sacred water was familiar, surrounded as we were by baptismal fonts, roadside saints bubbling water for thirsty travellers, and the Pellegrino bottling plant downriver. Privately, later, I noticed how much the duff jars were appealing in the same way 19th century naturalists' collections were, materials we had taken and displayed in a neat row of symmetrical glass. Our sketches were sometimes designs on behalf of the local cows or endangered orchids, but more often our ideas were serving human comforts.

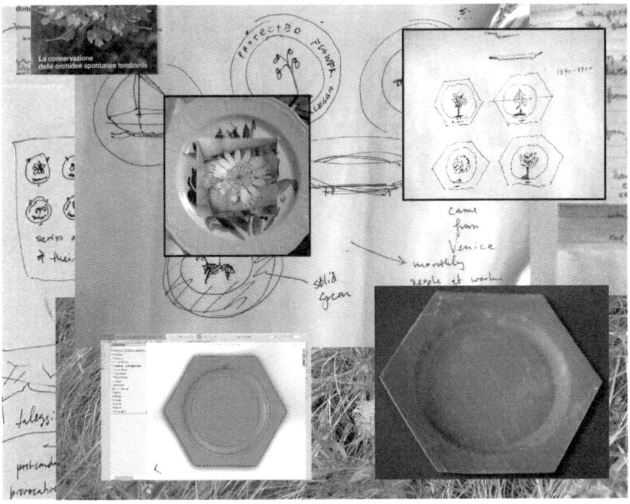

Sketches inspired by river samples.

An Ethnographic probe of cheese culture:

Another research method with a troublesome 19th century legacy is ethnography, implicit in violent colonial history. In design research, the ethnographic probe is sometimes called a cultural probe, or simply a diary study. I devised a diary study with the particular microbes, or cheese cultures, inhabiting three pieces of cheese, the rest of which went onto my breakfast sandwiches. I had brought with me small porcelain dishes of my own fabrication, and settled each cheese hunk in its own dish for a few weeks. Over time, this research activity generated the most pungent research results.

Diary study of microbial life on cheese.

As part of the NAHR design residency, we were given generous access to the cows, goats, farmers, cheesemakers, cheese tenders and storage caves. All these expert makers collaborated on producing the protected designation regional cheeses, most famously Taleggio as well as Stracchino and Brembana. As we toured through endlessly fascinating racks of aging cheeses, we saw some of the slabs and rounds had dramatic and ethereal molds growing inches into the air, and some of the storage caves were eye-wateringly acrid. A common topic on our tours was the differing standards across global markets regarding rawness of milk, hygiene standards in facilities and packaging, and variances in preferred whiteness of rind. We are learning much about our microbiomes and the interdependence and permeability between human and microbial beings.

As Heather Paxson writes in "Post-Pasteurian Cultures" (2008): "I introduce the notion of microbiopolitics to call attention to the fact that dissent over how to live with microorganisms reflects disagreement about how humans ought to live with one another. Microbiopolitics is one way to frame questions of food ethics and governance. ...Once cheese is recognized and valued as a living organism or microcosmic farm—once it becomes a microbiopolitical object—care of the cheese, care of the animals, care of the land, and care of the consuming self all must consider the microbe".

One evening in Sottochiesa, after a day of interviewing non-human persons in fields and barnyards, I discovered a tick had burrowed into a soft bit on my leg, behind my left knee. As I carefully tweezed this unsettling development into another jar, I considered Paxson's consuming self, and my consumed self: eating and being eaten. I recollected the Baltic-German biologist Jakob von Uexküll who was thinking of a tick as he developed the influential philosophical concept of *Umwelt*. *Umwelt* is a particular being's subjective perception of the surrounding word. I had read Heidegger, Derrida, Agamben and others who speculate on Uexküll's tick, its *Umwelt*, and that *Umwelt's* commensurability with our *Umwelt*. With my tick, my speculative research became embodied. There in the jar was a most visceral lesson in the Buddhist concept of the interdependence underlying all being. As Heesoon Bai writes in "Buddhism as a System of Education" (2015), "This concept points to the fact that nothing in life and the world is self-existing as an independent entity and everything is interdependent and interpenetrates. As such, everything is constantly re-assembled moment-by-moment and shows up as emergent phenomena". After an embarrassingly long ambivalence, I returned the tick to the field it might have come from, and perhaps parts of me live up there still.

Follow-up activities:

My research activities yielded many insights. A theming exercise generated three overarching categories. First, there were issues regarding communication and translation across species, especially the importance of listening. Second came considerations of scale among different types of agents. For example, at what pace might a design project proceed in order to collaborate with a sacred rock? A third theme concerned metabolic questions: the interconnected but distinct transitions involving consumption, taboo and death.

The post-human research activities also generated my own continuing design making. Prototypes in progress include porcelain artefacts designed to sprout, to grow living foods, and with augmented reality overlays and decals that might tell biographical stories of some of the non-human beings that I met. As a design teacher my post-human-centred research activities have helped me assign post-anthropocentric design briefs in class, and nudge students toward post-human design work such as empathy studies with cats, salmon and seals, design for yeasts and mycelia, and discussions around how to preserve dignity and agency in non-human design collaborations.

We designers, who rush toward problems with hopes of making the world a better place with our innovations and progress traps, can post-human-centred design research methods help us re-integrate science, culture and nature? Can the practice of radically participatory design activities challenge our hard-wired hierarchies of entitlement? Can absurd and repetitive, if poetic, research activities embody a theoretical shift toward relational bio-politics? Even if research notes aren't always in languages we can understand, there is wisdom in the responses of those ticks, watersheds, honeybees and ancient cheese cultures.

Thanks to Ilaria Mazzoleni, Alexandru Balasescu, NAHR board, fellow NAHR residents. Heesoon Bai, Julie Andreyev, Louise St. Pierre.

REFERENCES

- Bai, H. (2015). *Lotus Petals in the Snow: Voices of Canadian Buddhist Women.* Ottawa: Sumeru Press Inc.
- Latour, B. (1993). *We Have Never Been Modern.* Cambridge, MA: Harvard University Press.
- Paxson, H. (2008). Post-Pasteurian Cultures: The Microbiopolitics of Raw-Milk Cheese in the United States. *Cultural Anthropology*, 23(1), 15–47. doi:10.1111/j.1548-1360.2008.00002.x
- Sanders, L. & Stappers, P.J. (2013). *Convivial Toolbox: Generative Research for the Front End of Design.* Amsterdam: BIS Publishers.
- St. Pierre, L. (2015). Engaged Sustainable Design: Creating Moral Agency. *Nordes*, 1(6). Available at: http://nordes.org/opj/index.php/n13/article/view/381

SECTION II: SITTING

Sitting to open dialogue

In sitting, we invoke the timeless moment of the space of releasing a thought and of simply being open to ourselves, our life and our world. Immediately we can become aware of the body making contact with the floor, cushion, bench or chair. We may feel hardness or softness, sensations that change over time. Sitting names this section of the book as a starting point for close dialogue between humans and animals, plants, our ancestors and our futures. Here, sitting allows us to be gentle, patient, exploratory, imaginative in the dialogue. We profoundly question what it means to have voice and to participate: what can happen if humans and designers cease trying to control and programme multispecies interaction?

A featherlight visual essay by Li Jönsson and Tau Ulv Lenskjold opens the section on *Sitting*, highlighting the limits of human language in making sense of the natural world. By offering us photographs taken by birds on rainy winter days in Denmark, accompanied by a poem by Neil Bennun, Li and Tau invite us into a space where the boundaries between human and animal are blurred. The dualism of nature and culture is replaced by hybrids, and our human–nature relations come unsettled.

Drawing on a large scale transportation project in the Middle East, landscape designer Paola Sturla describes how the deterioration of material ecosystems are intertwined with a flattening of our perceptions of them through the limits of the visualisation technologies designers rely on. She proposes a landscape narration method, which engages with emotional and subjective experience of landscape as well as formal qualities, to support quality of life of humans and ecosystems.

Similar to Paola Sturla's notion of flattened perceptions of ecosystems, Jana Thierfelder problematises the removal of traces of the individual researcher and the 'creature as alive in nature' that takes place in biological science in the name of objectivity. She suggests designers collaborate with biologists to create 'thick descriptions' of nature that highlight its intrinsic value and show the interpretation processes that take place in science. These thicker images can evoke human entanglements with nature and the need to proceed with care.

The previous contributions in *Sitting* speak of how language and technology may create distance or may forge relations between nature and culture. Faith Kane, Huhana Smith, Rangi Te Kanawa, Tanya Maree

Ruka Te Miringa Te Rorarangi and Angela Kilford speak of how a plant, the harakeke, New Zealand flax and Māori tradition can be a focal point for remembering, learning, relating, stewarding, healing, imagining. Drawing on textile thinking, they are weaving a platform for learning from nature to benefit place and people.

Raune Frankjaer's piece centres around the potential of sympoeisis—*becoming with*—as she relays a project where it became necessary and possible for a design duo to work with agencies and preferences of plants to create an interactive installation. If previous texts in this part of the book ask for a language that can evoke humans' respect for and care of nature, here the focus is the possibility of communication between humans and non-humans. This in turn questions the nature of participation, and human preconception of human agency and voice as prerequisite to participate.

Voices of more than human species come back in London-based Anette Lundebye's essay as she explores how "to name is to value". She reflects on a design and art project where, in Linnaeus' footsteps, naming was used to evoke silent stakeholders—such as a bird, an insect, a polar bear—in order to make present the reality of the danger of their extinction, or of extinction itself. Building on this section's emerging theme of a need for many ways of languaging, Anette's piece calls for more than rational languages—to voice the precarity of the condition of ecosystems, and to voice the pain that comes from recognising what is being lost.

(12 April 2013) We are still gripped by winter and everything is brown. Though I heard and then saw a curlew the other day and I feel so much better for it.

Li Jönsson, Tau Ulv Lenskjold
Poem: 'Zoology' by Neil Bennun

Hybrids. Others/Selfies

In this visual essay we present to you a sequence of photos from the design research project Urban Animals & Us, accompanied by a reflection on the photo's inherent relations between humans and animals by the writer Neil Bennun. We encourage you to immerse yourselves into both writing and photos alike and contemplate your own reactions and responses.

Animals can change our outlook on and perception of the world. Human–animal relationships, no matter how superior and in control the former, involve a degree of adaption to the ways and wants of the latter, be it beast, bird or reptile. Such slight relinquishing of control can be put to experimental aesthetic use. In a study of 19th century Paris, Walter Benjamin (2002: 42) offers a description of how it was fashionable, for a time, for the flâneurs of the Parisian arcades to walk with a turtle on a leash. The turtle would determine the direction and slow the walking pace to allow the flâneur to dwell uninhibited and unproductively amidst the spectacles of modern urban life. The relationship between domesticated animal and man is turned upside-down as the flâneur is 'walked by the turtle'. But in the legacy of the flâneur as literary figure and vehicle for the critical examination of the conditions of modernity, turtle-walking remains little more then an exotic means to an end.

If we instead consider turtle-walking from the vantage point of our contemporary age of the Anthropocene, the turtle may have some significant things to teach us. Today, slowing down is no longer merely a subjective and aesthetic proposition: it outlines a general call to arms against the ever-accelerating speed of contemporary technoculture and its devastating effects on our shared worlds. Following the Belgian philosopher of science Isabelle Stengers' "Cosmopolitical Proposal" (2005), we may contend that "a slowing down of thought and action, enables the emergence and recognition of otherwise excluded human and nonhuman entities" (Farías, 2017: 35). Central to the concern of this proposal is the pronounced critique of modern society's dualistic distinction between nature and culture (Latour, 1993). To challenge this monumental dualism, then, is to hybridise human and non-humans—or, more specifically, as is the case here humans and animals—in ways that blur clear-cut distinctions through experimental accounts of interconnected networks and worlds.

Below is a series of seemingly random and chaotic snapshots from Urban Animals & Us. They were made during the speculative and participatory design experiment devised to investigate nature/cultural hybrids. The photographs were generated by birds through the experimental instrument InterFed[1] (fig. 1) that consists of two different photographic devices. One device is placed outdoor and equipped with a motion sensor instantly setting off a photo every time a bird (or other animal) interacts with it. Simultaneously, the instrument also activates the indoor device and captures a photo of life inside a building as it happens in that very instant. The simultaneous pair of photos are then juxtaposed and displayed side by side on a portable screen. The photos below thus represent three momentary instants of a continually produced visual essay of animal–human reciprocity, or a sort of selfie made by 'others' (than humans).

Most of those snapshots have never been seen by anyone, having flickered past on the portable screen when no one has payed attention, then to be silently saved onto computers, reposing as digital archives. However, while resting safely in the archives, they have continuously emerged and evoked discussions among us design researchers—how to make sense of them?

Over the years, we have come to realise that one problem with trying to make sense out of them has arisen from the fact that they have been unanimously caught up in academic discourse and as correlates of our research agendas. Language is such a central feature of being human, and the academic environment's advancement of knowledge is bound to complex discursive systems and technologies. Nonetheless, working with experimental design prototypes and investigating nature/cultural hybrids requires us to re-frame not only our understanding of the world but to experiment with representations of the worlds. Clearly, words rarely give images full justice and language, and (for the most part) may have been granted too much power of late (Barad, 2003). But language is also an inescapable fact of human life (yes, you're reading this, aren't you?). What

Figure 1: InterFed's three different devices. The outdoor and indoor device snaps a photo every time a bird interacts with it. The photos are then displayed side by side on the screen.

if we instead tried to engage the pictures below through an experiment in language?

What if we summoned the aesthetic qualities and sensibilities of language to make sense of the visual rhymes of bird-selfies created by InterFed?

In what follows, we have invited the writer Neil Bennun to engage with this material without any restrictions and only given the same rudimentary explanations provided here. This experiment, as you can see for yourself, offers a new story of nature/culture hybridity, resulting from a different writing practice and the author's employment of their own sensibilities, concerns, interests and skills.

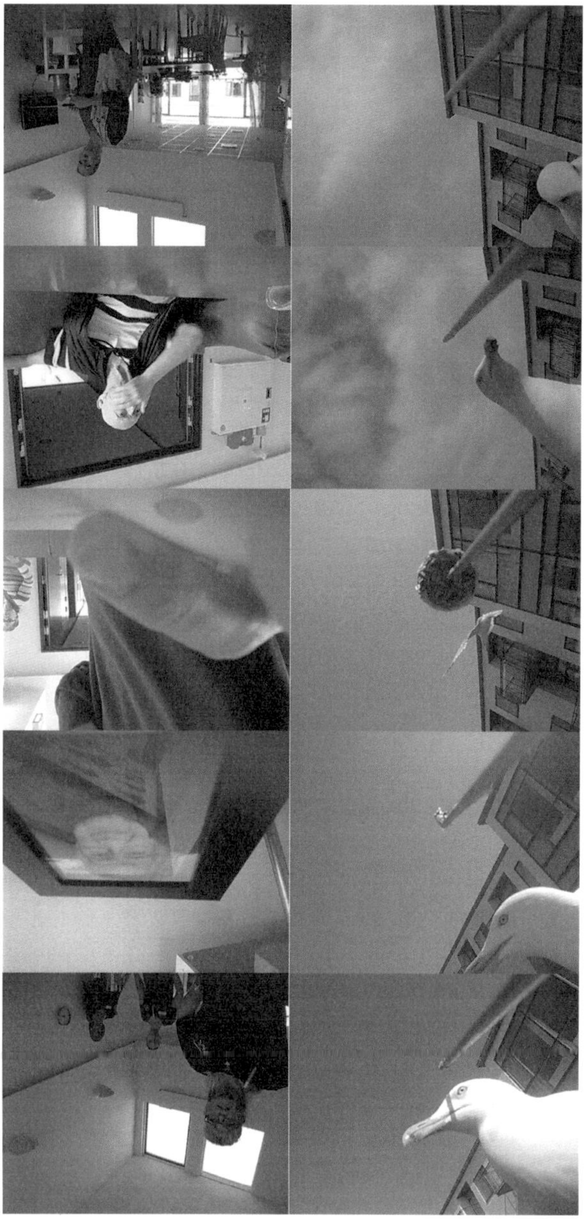

Inviting Neil Bennun to engage with the material.

1

Geology is the study of a complex
pressure attempting to
understand itself.

Cosmology is the study of a
complex pressure attempting
to understand itself.

Chronology is the study of the
manner in which this complex
pressure is maintained.

Anthropology is the study
of a dialect understood by
this complex pressure
in a single locale.

These words are written
at the coincidence
of zoological, geological
and industrial time.

To the left, sequence of photos. To the right, text response by Neil Bennun entitled "Zoology".

It is understood that the microplastics we might detect in the guts of these seabirds in Helsingør are as natural as the microbial fauna that will eventually evolve to digest them. It is understood that the isotopic signatures of anthropic development, now the study of geology, are equally natural. It is understood that the leak into the seas of heavy hydraulic lubricants from pipes and tankers, the final extinction of trilobites and sharks, the retreat of the glaciers, the strangulation of swans from the plastic collars of beer cans, primordial background radiation and the aggregations of heavy elements in parts of the cosmos too distant to be perceived, are all natural artefacts of an identical pressure.

Nothing in the universe is capable of inhumanity apart from human beings.

The fossil record reveals the ancestors of these birds to have been excellent geologists.

These birds have named the present geological period after us, in honour of our insight, in recognition of our role in the relief of a complex pressure rendering us fossil.

These birds are excellent photographers.

NOTES

1 InterFed formed part of the larger design research project Urban Animals & Us. The experimental prototype was deployed amidst senior citizens and birds at a care home and in the surrounding park area during a couple of particularly cold and rainy winter weeks in the Danish city of Elsinore.

REFERENCES

- Barad, K. (2003). Posthumanist Performativity: Toward an Understanding of How Matter Comes to Matter. *Signs: Journal of Women in Culture and Society, 28*(3), 801–831.
- Benjamin, W. (2002). *The Arcades Project*. trans. H. Eiland & K. McLaughlin. London: The Belknap Press.
- Farías, I. (2017). An Idiotic Catalyst: Accelerating the Slowing Down of Thinking and Action. *Cultural Anthropology, 32*(1), 35–41.
- Latour, B. (1993). *We Have Never Been Modern*. Cambridge, MA: Harvard University Press.
- Stengers, I. (2005). The Cosmopolitical Proposal. In B. Latour & P. Weibel (Eds.), *Making Things Public: Atmospheres of Democracy* (p. 994). Cambridge, MA: MIT Press.

(22 March 2016) I ran at lunch today and Canadian geese flew alongside me for about 10m. They are so creaky! Their wings and their gawping beaks.

Paola Sturla

Narrating the Impression

Having been a landscape designer in practice in the context of a large-scale transportation project in the Middle-East, I noticed the difficulty in retaining and communicating sensory features of my work. The continuous stressing on fast parametric modeling and image production steered me away from considering the sensory qualities of the environment and their relationships with my design, despite the fascinating desert context in which it is immersed. The sharp light on the dusty atmosphere surrounding the high-rise glass reflective façades, the bright glare that such light generates on the limestone built piazzas and the humid, dense air were not considered while designing the landscapes around the infrastructure. Later, when I ran into the "Design + Nature" call for contributions on design as a tool to tackle ecosystem deterioration, my mind went back to that work.

The assumption leading this essay is that the meaning of such ecosystem deterioration could expand beyond its necessary but limited materialist sense to include the flattening of the perceptive qualities of ecosystems. According to the Cambridge Dictionary, the word ecosystem means "all the plants, animals, and people, living in an area considered together with their environment as a system of relationships" (*Cambridge Dictionary*, 2018). The use of the word "environment" in the definition suggests that such relationships include the immaterial, perceptive ones. Here I refer to the concept of "*Umwelt*" defined in 1926 by Jakob von Uexküll, a German biologist who worked on animal behaviour and muscular physiology, and a founder of the biosemiotics field. *Umwelt* could be translated in English as "environment" or "nature"; however, it differs from the mere surroundings because of the active role of the subject in perceiving it, while being immersed in it. During an informal conversation on Uexküll's work I had with Dr. Lidia Gasperoni, Lecturer in aesthetics and space theory at the TU Berlin, it was highlighted that the original argument by Uexküll implied that the world of animals differs from the world of humans because of the human consciousness in grasping the subject position within the environment. However, advancement in research shows that we share this ability with animals (see for example McFarland, 1993). Therefore, I would argue for a more comprehensive concept of *Umwelt* that includes animals as active subjects. As a consequence, because the *Umwelt* is bounded into the physical form and materiality of the environment mediated by the subject's perception, its sensory qualities could enable what Immanuel Kant defined as the aesthetic experience (Kant, 1724[1900]), the subject's impression of the environment. The second assumption of this essay is that the aesthetic experience is an immaterial relationship linking the living inhabitants and the environment within the ecosystem. For this reason, while designing an environment, this aspect should be taken into account, together with more materialistic ecological strategies.

In my experience, the ability of grasping and imagining such impression was lost in the digital design process, probably due to the difficulty in measuring and parametrising such qualities in the binary language of design software. While designing environments, practitioners envision their products in 3D, 4D, 5D models, and visualise them through rendered images. Most of the time such representations focus on form and elements, and prevent them from retaining the complexity of information embedded in the *Umwelt*.

As a potential strategy to enrich the design of such environments, I propose the "landscape narration" method to vehicle such information, narrating and imagining the impression of the non-measurable qualities of both the existing and designed *Umwelt*. Hence, I rediscovered what I learned in professor John Stilgoe's class on "Adventure and Fantasy Simulation" at Harvard University on the immersion on what I am now calling *Umwelt*, and the narration as a way to communicate it, which has been a primary source of inspiration for this work. Let me clarify what I intend with narration, and why I have introduced the word "landscape".

Let's consider the user's aesthetic experience of the environment. I borrow the definition of aesthetic experience from Immanuel Kant, meaning the pleasure associated with the moment of judging something beautiful (Kant, 1790[1992]). In landscape design, to impress the user in this regard is one of the ultimate goals of a project, beyond its functional and standards-compliancy requirements. This statement could be easier to understand if we consider that the English word "landscape" comes from the German "Landschaft", which means both a region (a territory) and a scene (an individual visual perception of someone standing in that region).[1] This second definition underlines how the notion of landscape implies a mediation; there is no landscape without a subject experiencing it (and judging it). In the more specific case of a designed landscape, the mediation implies both the designer's interpretation and the user's experience of that landscape. Therefore, the attempt of situating the work "in nature", in the *Umwelt*, setting the ground for the aesthetic experience of the landscape, is a key challenge when approaching the drafting phase of the landscape design process.

How to bridge the gap?

Historically, both in literature and design sectorial press, scholars and writers have discussed how to render the impression of the environment. Nineteenth-century Romantic writers used hand drawing and written "word-painting" (Landow, 1985; Flaxman, 1998) as a tool to investigate the perceptive qualities of nature. They aimed not only to observe but to see the world in relation to personal feelings, therefore accounting for the mediation I mentioned before. This consciousness has been the result of an international debate among members of the picturesque movement. Landscape historian John Dixon Hunt, in his lecture on the reception of the visuals,[2] argues that John Ruskin understood that the mind, emotions and the act of perceiving mediate the visual stimulation in the process of drawing after comparing his sketch of the Pass of Faido in the Alps with the one by his friend William Turner (Hunt & Morabito, 2012). This consideration propelled him in reconsidering sketching as a tool to understand the world, adding the verbal extension of seeing, the *ekphrasis* (description) to capture the spirit of things (Hunt & Morabito, 2012). Following Dixon Hunt's argument, we discover that Ruskin intended the *ekphrasis* as a complex activity requiring abilities that transcend writing, enabling the artist to reproduce the personal impression of an object, not just its mere form (Hunt & Morabito, 2012).

Later, several scholars and professionals reflected on narration as a tool to imaginatively investigate perceptive characteristics of landscapes. I'll mention only a few of them as an example. Günther Vogt, a Swiss landscape architect, in his work "Landscape as a Cabinet of Curiosities" states "reading is much better for my imagination. For me, a text is much more open than a picture, [...] and imagination is always much stronger than reality, at least for me" (Vogt, Bornhauser & Kissling, 2015). Another example is the work of John Stilgoe on the "magic of Pennsylvania travel narratives" (Stilgoe, 2015a) in his studies on the relationship between landscape and images.

I mention the work of Professor John Stilgoe with emphasis: it is through his teaching work that I started questioning the role of design media in practice. In his class, after reading each week a science-fiction novel, and attending lectures on the evolution of science-fiction narration in relation to culture and fashion through the 20th and 21st centuries, the students discussed the role of imagination and perception in framing contemporary trends. The following is an example of what I call landscape narration from his work "What is Landscape?":

> *Landscape smells of the sea. Wading into the ocean on a summer day reveals the essence of landscape. Seaward lies deeper water, surf, riptides and undertow, ocean rollers, and an arc of horizon never still and mocking human control. Landward the beach glows in sunlight, interrupting the storms which sweep sand and dunes and smooth cobbles and boulders. Like the margin between the end of type on this page and the edge of the page itself, every beach is marginal, literally the marge, a limicole zone contested by wilderness and human order. With back to the sea, the wader sees sand and seawalls, as well as cottages and hotels and pathways that are beyond the reach of all but hurricanes and incremental sea level rise.* (Stilgoe, 2015b)

I want to share with the reader that, being a student in a design school, it was an unfamiliar situation for me to debate about landscape without delivering any visuals, which are the most common media in our profession. However, in my opinion, the focus on imagined worlds narration provided me with a method to sense the environment and to complement my design method. This action allowed me to enrich the content of my work and to consider aspects of landscape design that are not easily conceivable when working on parametric deliverables on the flat screen interface of a computer.

The fact of acknowledging the sensory qualities of the environment and its relationship with the subject is therefore a reason to experiment with landscape narration. However, so far my argument has been limited to the human centric, traditional approach to landscape design, in which the only subject involved in the mediation is the human. As I mentioned before, lately research has shown that animals perceive themselves, and they could be considered active perceivers in the game I am describing. As a consequence, in my opinion the argument could be pushed forward, getting out from the human-centred point of view to embrace a wider imaginative practice, narrating the *Umwelt* from different points of view, the ones of animals, insects, etc., leading to a design practice that acknowledges it. Beyond the contemporary literature on post-human approach to design, I believe that this exercise could help designers in embracing alternative perspectives, and deliver innovative strategies for ecosystem design.

After working on the landscape narration for the adventure and fantasy simulation class, I have applied what I have learned in my 2011 Master's degree thesis at Harvard, exploring the alternative point of view of bees in shaping a planning strategy for visual camouflage and ecosystem enhancement of the metropolitan region of Milan.

In conclusion, because of my personal experience in infrastructure design practice, I have noticed that the introduction of digital technologies as BIM software could gradually change the workflow, marginalising designers' contribution regarding situating their work in "nature". This condition leads to an ecosystem deterioration phenomenon if we consider an ecosystem as a set of complex relationships between the living subjects and the environment. This essay argues that such relationships include immaterial, perceptual ones. Therefore, I propose landscape narration as a tool to support the design process and enrich current practices by including impressions of landscapes to foster the quality of life of both humans and animals, the inhabitants of the ecosystem.

NOTES

1 For a broader definition of landscape see Stilgoe, 2015b, "What is Landscape?".

2 Published in Italian in the volume edited by Valerio Morabito, "Sette Lezioni sul Paesaggio".

REFERENCES

— *Cambridge Dictionary* (2018). Definition of 'Ecosystem', *Cambridge Dictionary*. Available at: https://dictionary.cambridge.org/us/dictionary/english/ecosystem [accessed 10 October 2018].

— Flaxman, R.L. (1998). *Victorian Word-painting and Narrative: Toward the Blending of Genres*. Ann Arbor, MI: UMI Research Press.

— Hunt, J.D. & Morabito, V. (2012). *Sette lezioni sul paesaggio*. Libria.

— Kant, I. (1724[1900]). *Critique of Pure Reason*. New York City, NY: The Colonial Press.

— Kant, I. (1790[1992]). *Critica del giudizio* (6th ed.). Roma; Bari: Laterza.

— Landow, G.P. (1985). *Ruskin*. Oxford: Oxford University Press.

— McFarland, D. (1993). *Animal Behaviour: Psychobiology, Ethology, and Evolution* (2nd ed.). Harlow, UK; New York: Longman Scientific and Technical; Wiley.

— Stilgoe, J.R. (2015a). *Landscape and Images*. Charlottesville, VI: University of Virginia Press.

— Stilgoe, J.R. (2015b). *What is Landscape?* Cambridge, MA: MIT Press.

— Uexküll, J. von (1926). *Theoretical Biology*. London: Kegan Paul, Trench, Trubner & Co. Ltd.

— Vogt, G., Bornhauser, R. & Kissling, T. (2015). *Landscape as a Cabinet of Curiosities: In Search of a Position*. Zurich: Lars Müller.

(21 November 2016) It is icy rain here blowing on northern air. And so dark.

Jana Thierfelder

Thick Description Through Visualisations *Towards New Representations of Nature*

The names of Charles Darwin, Alexander von Humboldt or Carl Linnaeus most commonly are associated with the study of nature. However, they are not only of interest from the perspective of science, but also for designers. I look to representational methods such as visual notation techniques in biological field work to reflect on the potential of the design discipline to help raise the awareness of human entanglement with nature and the necessity for its protection—not only theoretically but also as a practice. The artefacts of biological field work offer an interesting opportunity for design, science and the role of visualisations.

Early representatives of scientific research all used analogue notebooks that scientists would fill with visualisations, like illustrations, sketches, prototypes and organic artefacts among others during their work. These first documentations of nature that were often developed in collaboration with professional illustrators and artists, remind me of the standard employment of design tools. However, in disciplines outside of design, the researchers have to stick to the prevailing rules of objectivity that define how scientific observations should be represented, visually as well as in written words.[1] Despite the introduction of high-tech digital devices and the work with big data, many of today's biologists still employ the same analogue tools during their research. These visual tools usually fulfil two basic functional tasks in scientific research: they help to store and transport the observations researchers make during field trips (and thus create "immutable mobiles" as Bruno Latour would call them (1986)) and they help to communicate and share information among peers and the public.

The work with visual notations seems to play a key role in natural science knowledge production. In science, objectivity defines the aesthetics of visualisations. Consequently, a specific visual language is created that represents nature to its audience.[2] The technical-scientific objectivity paradigm the researchers have to meet defines how scientific results, i.e. papers and its visualisations appear, and how nature is perceived. A collaboration between design, with its expertise in visualisation practices, and the practice of large scale visualisations in natural sciences can be fruitful. Other means of visualisations can enable a "thick description of nature"[3] that helps to question the dominant paradigms and that can serve as an alternative representational format.

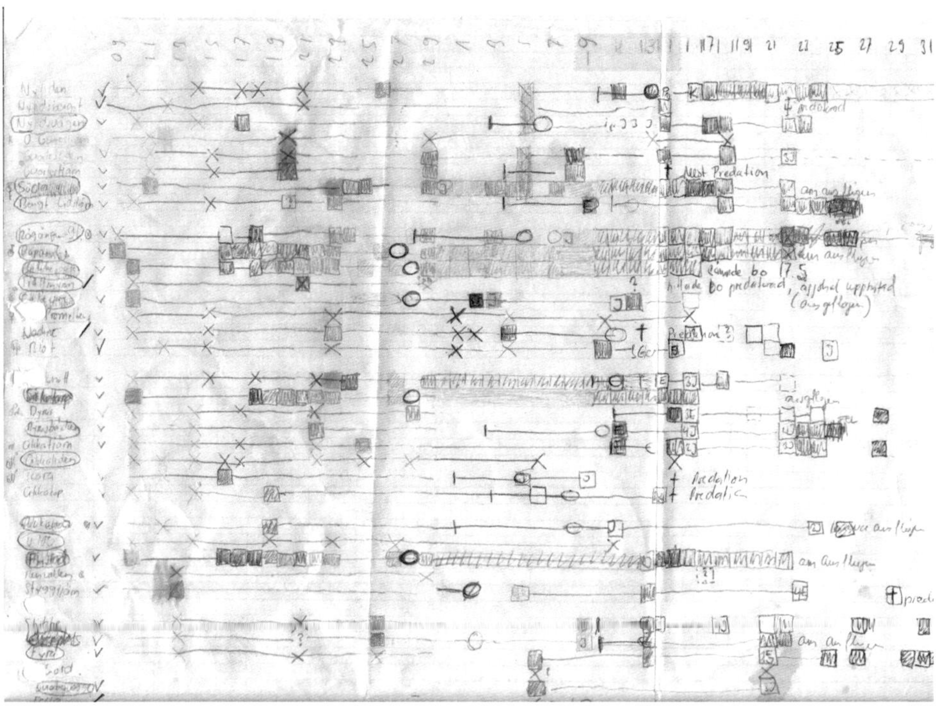

Figure 1: Handwritten documentation of the state of 34 nests of Siberian jays; with pen, colour pen and pencil on paper. Michael Griesser spring 2001. x = female is incubating, □ = young are hatched, † = nest predated or abandoned (see Griesser & Ekman, 2016).

Design practice as scientific practice

During field studies with Switzerland-based biologists, I studied the employment of visual notations and representational practices and processes that occur during the creation of scientific knowledge. In order to gain new insights about a research object, the researchers need to extract specific information about it. The main tool they employ is observation and documentation. In order to process the visual information that they

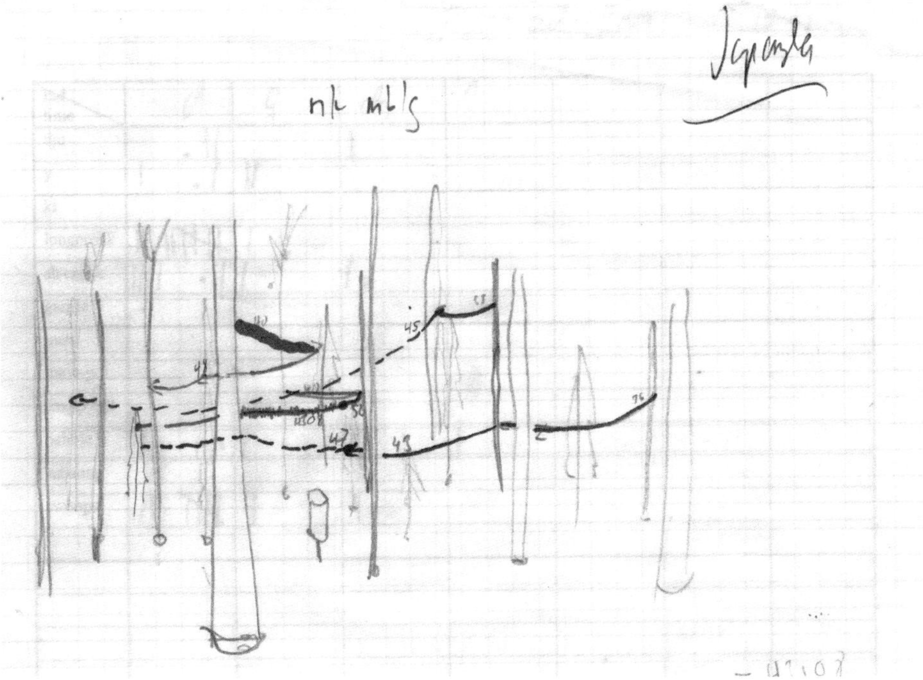

Figure 2: Hand sketches with pencil that help to visualise the movements of Siberian jays while exposed to an Ural owl model, autumn 1999 (see Griesser, 2009).

collect with their eyes, they have to translate it onto transportable means, such as field notebooks that enable them to store, fixate and transport it (fig. 1, fig. 2).

Aspects of this scientific work function similar to design processes and methods. This observation raised my initial interest into scientific representation practices and processes. I describe the diverse outcomes of scientific research as design artefacts, and consequently as results of a design process. This allows me to draw relations between the two fields, science and design, and leads to an investigation into the opportunities that lie in a cooperation of scientific and design practices.

Both processes usually start with an idea, followed by drafts and layouts that are then realised. While in science the output usually is a research paper, in design it is some sort of an object. In both cases, visual notation techniques like drawing and drafting support the thinking process and work as exploration into the subject. This explorative process is shaped by trial and error, by prototyping and testing phases. Visual notations and representations usually follow a communicative aim, formal-aesthetical rules and fulfil certain functionalities, i.e. transporting and storing data, and processing raw data into scientific facts. They thus help to channel and communicate information. In both disciplines, visual skills are essential to work with often microscopic aspects of the world. Finally, designers and biologists are experts when working with vision, thereby understanding vision as a specific skill, immanent to the discipline, a "skilled vision" (Grasseni, 2011).

These similarities in the daily practice of design and science help to understand the relations between them. They stress why designers should get involved with scientific processes to help to develop new representations of nature and thus our environment. Rather than affirming the status quo, designers and scientists can unite to raise new questions and thus create change.

The information field notes carry

There is a striking difference between primary raw scientific data that result directly from field work and the translated representations in the final results. The raw field notes give deep insights into the research processes and its conditions. The hand-written notes, with the sketches and illustrations in the notebooks, reflect the human beings behind them and make the research object aesthetically and emotionally visible. Each individual notebook has a certain character that offers a notion of the biologists' passion for their subject. They give an impression about the life of, say, birds in nature. Finally, they serve the biologists as a highly effective memo technique, allowing them to precisely memorise almost each step and its conditions within the field study. However, these visual raw data have to undergo an objectifying transformation process to stick to the scientific rules of objectivity. This filters out all traces that give insight into the aesthetic aspects of research that lead back to the researchers as individual human beings, and the creatures as alive in nature.

All visual notations that were produced during field work usually are translated into Excel sheets (fig. 3) and from there into a scientific graph with the help of programming applications, reducing the information to one seemingly clear piece of information about the research subject. The handwriting of the researchers, the visibility of their diverse notation skills, the conditions of the field work and their motives that were visible within the field notes are erased by digital devices and computers. Only then do the visual representations count as a scientifically valid result. During this process the researchers have to make themselves invisible from their own work in favour of the representation of an objective research collective and the unification and simplification of their data set.

Yet, what happens to the notations and their value as raw data during this process? As described earlier, the first drawings show a lively, vivid, unique picture of nature. They convey the emotions the work triggers in the biologists. However, the final graphs seem to freeze only one aspect

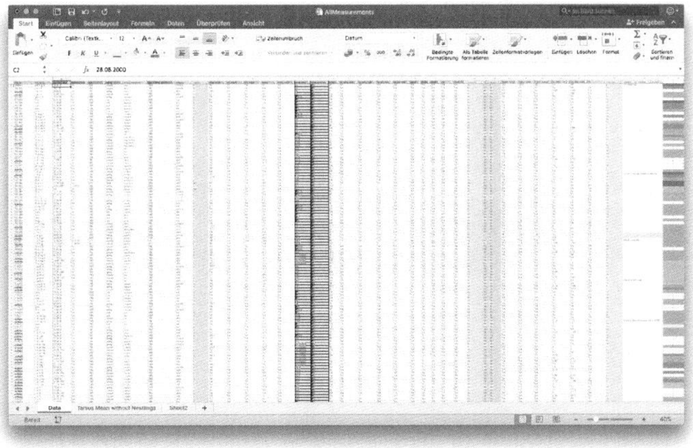

Figure 3: Raw data plots in an Excel sheet, showing all data collected during a study with Siberian Jays.

about the research into an abstract scientific fact. What the researchers had actually seen with their bare eyes and literally collected with their hands on paper during field work has lost its value and has to be replaced by a so called objective visual language, i.e. graphs and diagrams. By doing so, the researchers aim to draw a specific picture of nature that represents their outcomes as distinct results of a clearly planned, linear research process. No longer are visualisations employed for storing and transporting information, nor is the iteration that shapes the process visible. The representations now work as a visual proof for a research result that conveys a clear, unmistakable and strong message; the communication of their scientific findings. The difference between the primary notations and the final visualisations thus lies in the information they carry and the functions they fulfil.

As a consequence, the images about nature that circulate in the public follow a representation according to the prevailing technical-scientific objectivity paradigm. However, these representations have lost a lot of important information and have only little to do with the actual research conditions and researched objects.

New representations of nature

From a critical design perspective, there is a high awareness that images have their own policies and thus impact how scientific peers and lay people perceive their environment through them.[4] The way in which nature is visually represented to the public impacts people's behaviour towards nature. These prevailing representational practices and their consequences, thus, have to be reflected on and critically rethought with the help of designers.

It might seem obvious that we could simply publish the sketches and field notes along with the research outcomes. But this information would overwhelm the public. This is where designers can engage in the situation, collaborating with biologists to create a new thick description of nature. Rather than representing nature as something that can easily be dealt with by employing the correct scientific methods, new images of nature should point out its intrinsic value, independent from fulfilling cultural, innovative, and scientific interests. Designers and scientists should work together to bring new images into circulation.

This thick description of nature has the ability to make the transformation and filter processes transparent and accessible to the public, i.e. as infographics. Publishing qualitative along with quantitative facts further helps to draw the attention to the different functions visualisations can serve, thereby questioning the prevailing modes of representation. Rather than limiting the thought space of a scientific publication to one conclusion, additional visual material like field notes can help to open the space for imagination, allowing the audience to connect to nature in a much more intuitive and emotional manner, thereby making the iteration of the thought process of the biologists and of the research process accessible. This supports the description of nature beyond the prevailing categories, helping to move away from the technical-scientific objectivity paradigm towards a visual language that represents plurality, diversity and multiplicity and thus constructs new meaning about nature, along with scientific facts.

Conclusion

Design can help to express what scientists cannot within their practice due to its rules and regulations. When confronting challenging ecological crises such as global warming, shortage of water or the extinction of living organisms it is essential to rethink the man–nature relationship. Sticking to the prevailing roles of objectification as a visual representational practice is no such way. Images interact with humans. By providing better images of nature, the scope of human action can be triggered. Design, as a transdisciplinary discipline that is able to handle complexity and make it accessible, has the ability to engage with diverse stakeholders and create new interaction. Design tools can be employed to raise the awareness of human entanglement with nature and the necessity for its protection—not only theoretically but also as a practice. With the help of design tools, the access to knowledge can be made more democratic, engage a wider audience and facilitate change. This is one opportunity to overcome disciplinary boundaries as well as the dichotomy of nature and culture.

All images have been provided by Michael Griesser, head of an evolutionary biologist research project, that has conducted studies on a bird species, called Siberian Jays, for over 60 years.

NOTES

1 Lorraine Daston and Peter Galison (2010) outlined in their publication Objectivity how the rules of scientific representation have changed over the centuries and how it affected the construction of the researching subject within the visualisations.

2 In the context of a development towards an overlap between disciplines, such as art and science, and a growing demand for scientific communication, the public is defined by a heterogeneous group of various representatives that consumes scientific results in professional journals as well as in popular media, such as the daily press.

3 Glifford Geertz (1973) developed the term of a "thick description", that was introduced by the philosopher Gilbert Ryle, further and used it to describe an ethnographic research method. With this term, he aimed to not only describe the observations themselves but also in their context and their further relations.

4 Compare e.g. Silverman, Kaja. (1996). *The Threshold of the Visible World*. Routledge.

REFERENCES

— Daston, L. & Galison, P. (2010). *Objectivity*. Cambridge, MA: Zone Books.

— Geertz, C. (1973). Thick Description: Toward an Interpretive Theory of Culture. In *The Interpretation of Cultures: Selected Essays* (pp. 3–30). New York: Basic Books.

— Grasseni, C. (2011). Skilled Visions: Toward an Ecology of Visual Inscriptions. In M. Banks & J. Rubi (Eds.), *Made to be Seen: Perspectives on the History of Visual Anthropology* (pp. 19–44). Chicago, IL and London: University of Chicago Press.

— Griesser, M. (2009). Mobbing Calls Signal Predator Category in a Kin Group-Living Bird Species. *Proceedings of the Royal Society of London B: Biological Sciences*, *276*(1669), 2887–2892. doi: 10.1098/rspb.2009.0551

— Griesser, M. & Ekman, J. (2016). Siberian Jays: Delayed Dispersal in Absence of Cooperative Breeding. In W.D. Koenig & J.L. Dickinson (Eds.), *Cooperative Breeding in Vertebrates: Studies of Ecology, Evolution, and Behavior* (pp. 6–18). Cambridge: Cambridge University Press. doi: 10.1017/CBO9781107338357.002

— Latour, B. (1986). Visualisation and Cognition: Drawing Things Together. In H. Kuklich (Ed.), *Knowledge and Society Studies in the Sociology of Culture Past and Present* (vol. 6, pp. 1–40). Stamford, CT: Jai Press.

— Silverman, K. (1996). *The Threshold of the Visible World*. New York: Routledge.

Faith Kane, Huhana Smith, Rangi Te Kanawa, Tanya Maree Ruka Te Miringa Te Rorarangi and Angela Kilford

Learning from Harakeke *Towards a Network for Textile Design in Aotearoa New Zealand*

Learning from harakeke

In the process of coming together around harakeke (*Phormium tenax* or New Zealand flax), one of Aotearoa New Zealand's most distinctive native plants (fig. 1), the authors have spent time in korero (discussion) and have reflected upon our experiences of working with this plant, some over just a few months and others over decades. Here we share, in the first person, the significance of harakeke to us, and how it is working to shape our thinking and research towards a renewed harakeke textiles industry in New Zealand. This evolving narrative weaves together emergent themes of connection, identity, memory, time and healing as significant to each author. We propose that these themes will inform future endeavours in this field, creating a network of ideas that might inform future design-led research in this space.

In considering the importance of identity first within this emerging narrative, it seems pertinent to note the background of each author in order to provide a clear context for the accounts that follow. Rangi Te Kanawa, Tanya Maree Te Miringa Te Rorarangi Ruka, Angela Kilford and Huhana Smith all whakapapa or have indigenous genealogy to Māori ancestors of Aotearoa New Zealand: Rangi Te Kanawa belongs to Ngati

Kinohaku, and Ngati Maniapoto, centrally located in the North Island; Tanya Maree Te Miringa Te Rorarangi Ruka is of Ngati Pakau, Nga Puhi, Waitaha tribal descent, her tribal area is located in the Hokianga which is the far north of the North Island; Angela Kilford belongs to Ngāti Porou and Ngāti Kahungunu from the east coast of the North Island; and Huhana Smith has genealogical ties to Ngāti Tukorehe and Ngāti Raukawa ki te Tonga of the Kuku Horowhenua and Tapapa, Te Kaokaoroa o Patetere of Waikato regions. Faith Kane is a recent immigrant to New Zealand from the Midlands in the UK and has lived in Wellington since 2016.

Figure 1: Harakeke Plant.

Harakeke signifies identity — Rangi Te Kanawa

Identity is knowing your whakapapa (geneology) and the culture that is inherited, to this my lineage passes on the knowledge of harakeke. As seen in my wharetupuna (ancestral house), the carving of Rawinia — holding fast a skein of muka (the fibre of harakeke). Traditional Māori weaving is my lifetime experience. The strength and lustre of harakeke provide a wide platform to weave and construct functional items. Processing the leaf to fibre, the fibre to threads and dyeing the threads brings an awareness of te taioa (the natural world), which is integral to the concept of whakapapa. Identifying and developing sustainable practices in gathering the muka fibre will ensure a healthy resource for future textile innovation.

Also inherited, is the practice of traditional Māori weaving that affords me the role as conservator of stabilising taonga (treasured artefacts of traditional knowledge). In this role the makeup and cultural content of taonga is revealed for future generations. Such is my commitment to weave, research, stabilise and explore potential applications for our modern world with harakeke, that any contribution to advance this taonga is with privilege.

Harakeke sustains relationships through time — Tanya Maree Te Miringa Te Rorarangi Ruka

Due to my recent research I have become kaitiaki (guardian) of a kaitaka (cloak) that is woven solely from harakeke fibres circa 1860s. This piece not only connects me to my ancestry but also to the whenua (land) of that time. Kaitiakitanga (stewardship) of the whenua informs my art practice. As a contemporary Māori arts practitioner, working with harakeke as a sustainable solution to contemporary environmental issues has helped to grow a deeper connection to my culture, ancestry and practice.

Previously I created experiential installations within the gallery space by digitally weaving and projecting video of landscape imagery and soundscapes. Introducing harakeke fibre research has developed the practical aspect of kaitiakitanga that I was looking for within the work. This is based on the combined beneficial properties of the plant for the land and health benefits for the people.

The women of my tribe have woven together the fibres of harakeke for generations. My tupuna (ancestors) 'encoded' knowledge and stories into the textiles. This was done by observing correct tīkanga (methodology and procedures) in terms of when and where the plant was grown, gathered and processed, right through to the final woven piece of clothing. Exploring the possibilities of harakeke regarding new sustainable textiles and materials, weaves together a pattern of connection that spans past, present and future: 'I nga waa o mua' (looking back to the past to inform our future).

Harakeke holds memory of the land — Angela Kilford

In recent projects I have gone to great lengths to draw attention to the way in which whakapapa (geneology), as a system of knowing and looking at the natural world, reveals unexpected connections between people and their surroundings. Whakapapa relates all living things to each other, and reminds us of our relationship to papatuanuku (Earth Mother). In Māori cosmology all things stem from one set of primal parents linking people to plants and specifically here, harakeke.

In recent korero (discussion) with Tanya and Faith, Tanya brought along her special kaitaka (woven cloak of muka harakeke). She had been to visit Rangi already to talk about how to preserve this taonga (treasured artefacts of traditional knowledge) before taking it back to her whanau (extended family) and iwi (extended kinship group). As we studied the kaitaka Tanya observed that it was not only a fine specimen of weaving but due to its material quality, also carried the memory of the landscape from the time that it was made — when the plant and culture of making flourished together. This profound observation brought together what I knew already about whakapapa and reinforced my understanding of a Māori valuing of whenua.

Harakeka heals — Huhana Smith

Our ngāti tukorehe or iwi-based investigations into the potential of a revitalised harakeke industry have always centred around consolidated agency to not only rebuild connections to harakeke and its wide customary knowledge, but also to highlight its many healing properties for the sake of freshwater and the range of environmental degradation faced in places such as Kuku, Horowhenua. Since 1995 many a marae (ancestral traditional Māori meeting house) based discussion has taken place about futures with harakeke and its ability to rehabilitate lands, people, waterways and stock. Specific wānanga (intensive learning opportunities) in 2005 called for this taonga species (as a versatile natural resource) to be revitalised for its sustainability potential. Since this time and in closely following the work of Rangi Te Kanawa, as a leader in the harakeke textile space, the commitment to develop a sustainable and beneficial industry for our hapū (kinship groups) in Kuku, Horowhenua remains strong. The intent remains to build upon ancestral legacies and historic interactions with harakeke.

Combining science and innovation goes towards unlocking the potential of Māori knowledge, people and resources for the benefit of New Zealand. New knowledge around old industries should be internally meaningful to culture and hold relevance to issues for a contemporary Māori world (Royal, 2004). Harakeke not only creates possibilities for new enterprises within renewed industries, but also offers potential employment opportunities for Māori communities in the context of Māori lives unfolding today. In noting the harakeke plant as a whole-plant healer or a whole-of-system healer (Smith, 2007), the potential of indigenous agency and innovation contributes to the possibility of economic growth through distinctive research with, and development of, mātauranga Māori (Māori knowledge) around harakeke. The beneficial uses envisioned for harakeke include economically sustaining marae with raw muka (the fibre of harakeke) product for renewed harakeke textiles; for sustaining lands, waterways and improving farm animal health for eventual human use or consumption.

Harakeke connects people and place — Faith Kane

My desire to pursue textile design research for sustainability through a place-based approach (Kane, 2017) found a sharp focus on arriving in a new place — Aotearoa New Zealand — as an immigrant from the UK in late 2016. With an interest in European flax, learning about harakeke seemed a natural starting point from which to consider local resource flows for textiles. As I have slowly begun to engage with this plant as a starting point to explore local textile expressions and potentials, it has proven to be truly life-giving by connecting me to people and places in a deep and rich way, which is difficult to portray in words. Feelings of vulnerability in venturing into a new cultural space and a sense of uncertainty about how to engage respectfully and with integrity ensued. But, through a common vision for locally rooted textiles and knowledge of ways of working with textile fibres and processes that are common across cultures (Philpott, 2011; Igoe, 2013; Philpott & Kane, 2016), I have been kindly welcomed into the harakeke space.

In my role as a researcher in the School of Design at Massey University New Zealand (MU), I have had the privilege of learning of and from traditional Māori values, practices and knowledge that form the foundations for future textile design innovation, through being introduced to key people working in this space by Huhana Smith (MU) and Eve Kawana Brown (MU). In particular, I have learnt from the wisdom and insights gained over decades by the other authors contributing here including Rangi Te Kanawa who leads in this space, and further through precious time spent with artists Angela Kilford and Tanya Maree Ruka Te Miringa Te Rorarangi, in which we have shared our research practices and spent time in the landscape together. The possibility to build from these foundations in a way that enables all to retain their mana (authority) through carefully considered collaboration or manaakitanga (reciprocity), as reflected upon above, has been graciously shared by Huhana Smith's interdisciplinary work that focuses on environmental regeneration strategies for New Zealand.

Creating a network of knowledge

Through the wisdom gained in our korero (discussions) we are exploring an approach to future harakeke textile design in Aotearoa New Zealand that is grounded in mātauranga Māori (Māori knowledge) and adheres to tīkanga Māori (methodology and procedures) in order to provide a platform for engagement where all parties retain their mana (authority) through manaakitanga (reciprocity) and carefully considered collaboration. We perceive that this may interlace with other forms of knowledge that are often drawn upon through making textile such as shared tacit knowledge of textile construction, textile engineering, materials science, chemistry, soft logics and materials experience. Through an intention to begin to articulate our approach to future textiles design in this space, we are drawing on recent notions of 'textile thinking' (Philpott, 2011; Igoe, 2013; Philpott & Kane, 2016) that capture such networks of knowledge. Here we suggest new intersections based on the wisdom of the harakeke plant in regard to connection, identity, memory, time and healing (fig. 2). The network shown is indicative of further research planned with an extended group of researchers, who will continue to gather around the harakeke plant to direct textile development in Aotearoa New Zealand that actively learns from nature to benefit both people and place.

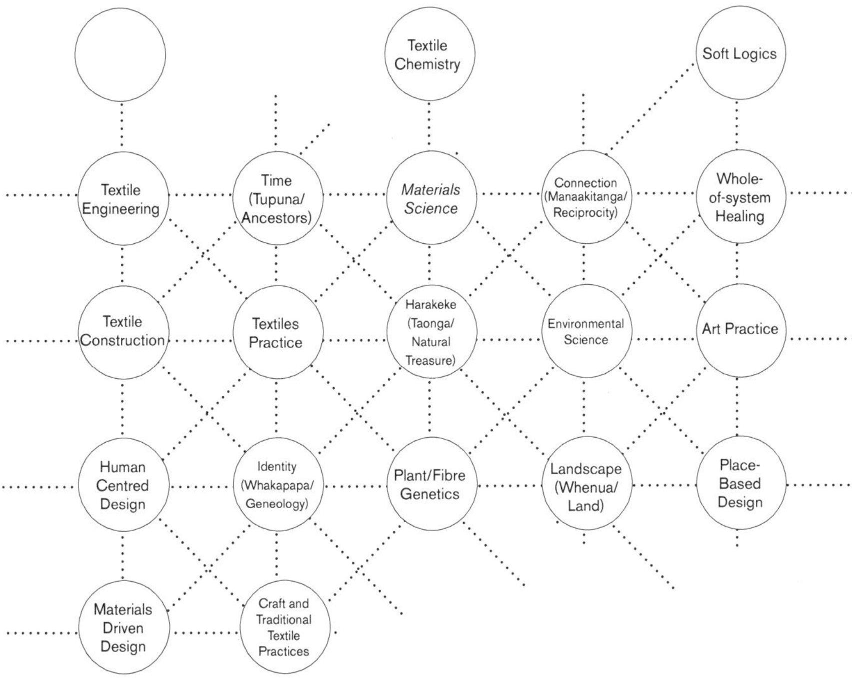

Figure 2: Evolving network of knowledge for future harakeke textile design emerging from korero (discussion) between Rangi Te Kanawa, Faith Kane (Massey University New Zealand (MU)) Huhana Smith (MU), Tanya Maree Ruka Te Miringa Te Rorarangi, Angela Kilford and an extended research group including Katarina Tawiri (Landcare Research New Zealand), Xiaowen Yuan (MU), Vaughan Simmonds (MU), Chris Anderson (MU), and informed by Kane (2017), Philpott and Kane (2016), Igoe (2013) and Philpott (2011). With reference to Materials Driven Design (MDD) (Karana et al., 2015), Place-Based approaches to Design (Norton & Hannon, 1997) and Whole-of-System Healing (Smith, 2007).

REFERENCES

- Harakeke and Wharariki Information Sheet (no date). *Landcare Research*, viewed 18 October 2017. https://www.landcareresearch.co.nz/resources/collections/harakeke
- Igoe, E. (2013) *In Textasis: Matrixial Narratives of Textile Design*. Digital PhD Thesis, London, Royal College of Art. http://researchonline.rca.ac.uk/1646/1/IGOE%20Digital%20Thesis.pdf
- Kane, F. (2017). Materials Design for Sustainability: Connecting with Local Resource Flows Through the Development of Flax-based Composites. *Craft Research*, 8(1), 101–118.
- Karana, E., Barati, B., Rognoli, V. & Zeeuw van der Laan, A. (2015). Material Drive Design (MDD): A Method to Design for Material Experiences. *International Journal of Design*, 9(2).
- Norton, B.G. & Hannon, B. (1997). Environmental Values: A Placed-based Theory. *Environmental Ethics*, 19(3), 227–245. www.cept.unt.edu/vol/vol19.html [accessed 24 October 2016].
- Philpott, R. (2011). *Structural Textiles: Adaptable Form and Surface in Three Dimensions*. PhD Thesis, London Royal College of Art.
- Philpott, R. & Kane, F. (2016). Textile Thinking: A Flexible Connective Strategy for Concept Generation and Problem Solving in Interdisciplinary Contexts. In T. Marchand (Ed.), *Craft Work as Problem Solving: Ethnographic Studies of Design and Making*. Abingdon, UK: Ashgate.
- Royal, C.T.A. (2004). *Mātauranga Māori and Museum Practice*. A discussion paper prepared for the Museum of New Zealand Te Papa Tongarewa, 26.
- Smith, S.M. (2007). *Hei Whenua Ora: Hapū and Iwi Approaches for Reinstating Valued Ecosystems within Cultural Landscape*. Unpublished Doctoral thesis, Te Pūtahi ā Toi, Massey University, Palmerston North.

(12 December 2016) Hoping you're fine in all this dark.

Raune Frankjaer

Becoming-with Vegetal: Sympoietic Design Practice with Plant Partners

In this chapter, I describe the profound changes in the relationship between us, two human designers, and various plant participants, which took place during the design process of a plant-controlled, interactive installation. Furthermore, I examine the nature of participation in relation to vegetal being and promote the notion of sympoiesis as an approach to design with nonhuman others.

We, the human designers in the project, both have degrees in media design, one working within 3D animation and the other pursuing a PhD in digital design. The plant participant, a Calathea has its ancestral roots in the Amazonian jungle but lives with the first designer and her partner in Germany. We chose this particular Calathea for the project due to its striking aesthetics and the physical qualities of its broad leaves, suitable for attaching the sensors needed for the installation's computational system. The objective of the project was to expose existing misconceptions around the nature of vegetal being as non-sentient and forge deeper, more meaningful relationships between plants and humans.

The installation uses data from the plant's electrical signalling pathways to control the colour-animation of an illuminated fabric-based artefact. Plants live sophisticated lives and possess an acute sensitivity to environmental changes and stimuli (Chamovitz, 2012; Karban, 2008). Due to their high responsiveness to touch, humans and plants can engage and jointly determine the artefact's visual appearance. This playful interaction between humans and plants prompts a re-evaluation of the human perception of the vegetal environment as passive automata and challenges prevailing ontologies that frame the natural environment as a resource existing merely to be exploited. The installation has been presented in Denmark, Germany, Spain and Colombia.

Designing with plant partners:

As the installation revolves around plants, we from the onset acknowledged them as participants, much in the same category as the humans who would participate in the finished installation, i.e. as someone to design *for*. However, during the design process the plants' behaviours challenged the structure of this relationship and forced us to acknowledge them as co-designers, i.e. as someone we needed to design *with*.

Describing plants as participatory entities may initially seem peculiar as participation is commonly understood as describing a distinct act by a conscious agent. However, what exactly constitutes participation is not as clear as often assumed. A 2015 study of Participatory Design, a design approach aimed at democratising design processes, based on the participation of various stakeholders and users, notes the lack of clarity and multitude of both explicit and implicit definitions of the nature of participation in contemporary Participatory Design practice (Halskov & Hansen, 2015). Several distinctions have been made with regard to the kinds of participants: end-users, stakeholders or decision-makers; the type of participation: direct participation by end-users, indirect participation by user representatives, or imagined representation, e.g. through personas or advocacy (Bossen et al., 2010; Grudin & Pruitt, 2002; Mambrey et al., 1998). Hence, participants are not merely explicit users but include stakeholders, i.e. those who do not use a design but nevertheless are affected by it. Nonhumans, e.g. animals, plants and even abiotic entities are often severely affected by human (design) decisions. Currently we are witnessing planet wide collapses of various ecosystems, a major extinction event, ongoing pollution and environmental destruction, all caused by the way we as humans have chosen to design and construct our societies (at least the dominant ones). Consequently, plants are clearly stakeholders in human design processes. Moreover, participation does not necessitate direct action but may be representative or even imagined, as when using personas in the development process. Andersen et al., drawing on actor-network theory, posit "participation as a 'matter of concern', something inherently unsettled, to be investigated and explicated in every design project" as opposed to a 'matter of fact' something binary that either is or isn't (Andersen et al., 2015).

For us the 'trouble' started when writing the programme for the micro-controller to map the electric signals emitted by the plant onto the optical fibres of the fabric as varying colour-patterns. As is often the case before premiering an exhibit, we were working tightly onto a deadline but were assuming a rather easy job. The algorithm was not too challenging and we had some code we could recycle from a previous project. We worked on the physical artefact construction and the weave during the day and late in the afternoon connected the plant to the fabric still on the loom, as pictured in fig. 1.

Figure 1: The photo-optical fabric on the loom with the connected plant participant in the background.

As expected it worked without any issues. The plant responded nicely to a gentle stroke of its leaves, and the colours were displaying beautifully on the fabric. We left for the night.

In the morning nothing worked. It seemed we were no longer picking up the signal from the plant. Assuming a fault in the system we first set out to troubleshoot the hardware, all the connections, the chips, the sensor-pads, the controller. Nothing. Aware of the futility of looking for bugs in the code — after all it was working just fine eight hours ago — but at a loss of what else to do, we started rewriting it. Suddenly, we could read the signal again and although uncertain of what exactly had caused the change, we were happy and returned to work on the qualities of the colour-animation, and the physical construction of the artefacts for the rest of the day.

Yet, upon returning the following morning, we again found nothing worked and commenced trouble shooting. After some time fiddling around unexplainably it was fine. Next morning, again — nothing. Only then did we realise there were no bugs in the code or faulty hardware, this plant was simply not responsive before around 11 am. Like us, it had a circadian rhythm and this was just not a morning plant. We subsequently moved our working hours to cater to the plants' preferences and were able to finalise the exhibit without further issues.

Next, we needed to acquire bigger plants for use in the exhibition and teamed up with a local nursery. Armed with laptop and sensors we made our way through the gardens searching for responsive outdoor plants, an endeavour which proved rather problematic due to the cold early March weather leaving the majority of plants still in winter dormancy. After some time, we settled for the five most responsive plants and loaded them into the back of the van, which would house them over the next five days of the event. As the exhibition was taking place in a public space during evenings we had to lock them and the artefacts up every night. Since we were at the exhibition site daily interacting with several thousand visitors and thereafter had to take everything down late at night, we were planning on sleeping in and relaxing during the day, getting to the site at dusk to set up.

The plants were not pleased.

On the second evening, their responsiveness was remarkably lowered, and so on the third day we decided we had to get to the site early in the afternoon to get the plants out of the dark, cold van and into the sunshine. It helped. And once again we changed our schedule to accommodate the needs of the plants, subsequently taking them out sunbathing every day, much like a dog owner walks the dog. What was becoming increasingly clear was that the power relations between us and the plants had shifted. We were no longer in control of the exhibit, the plants were. We had gone from our elevated position as creators to caretakers with one single objective: keep the plants happy. The experience was profoundly humbling and has deeply impacted our way of designing as well as our sense of being in the world. The intellectual abstraction, which signified our approach at the beginning of the design process, gave way to a carefully questioning empathy towards our vegetal partners. To achieve our goals, i.e. relaying plant sentience to a wider public outside of academia, we had to start thinking with the plants and develop a plant-based understanding. How does the plant relate? How does it feel to be a plant in this situation? What would the plants like? We, as humans could for obvious reasons not think as plants, but we could question how we, as caretakers, could create acceptable conditions for our plant participants.

A common argument is that plants are not capable of participation, as — even if we were to accept they have some sort of preferences or even intent — we cannot possibly know since we cannot communicate with them. Yet, one of the very basic qualities that signifies the skilled designer are ways of knowing if a design is working well for a user group without having to directly ask them or make them fill out questionnaires. Is it not rather so, that by assuming participation to lie purely with the (human) individual, able to consciously express his or her wants and needs in a human language, most of the participation goes unnoticed and is lost? Andersen et al. (2015) state: "When participation is a matter of concern, it is not possible a priori to consider some forms of participation to be more authentic or ethically superior to others. And, neither is it possible to establish a dichotomy between participation and non-participation. Rather, participation is configured and may come in many forms". At the exhibition the dynamics of the relationship between us, the designer-researchers, and the plants was extremely modified by the plants' responses to the situation. From the initial event design catering to our preferences, such as sleeping in and just getting to the site by nightfall, it quickly became clear that we had to let the plants dictate the temporal and spatial parameters of the installation. We ourselves had fallen into the trap we were trying to oppose, imposing our will and instrumentalising the plants by disregarding their needs, in this case for some time in the sun. Only by submitting to the agency of the plants, allowing them decisional power and by cooperatively designing the event, could we achieve positive results.

Sympoietic making:

One of the signifying traits of the skilled designer is a highly developed empathic sense, a way of 'designerly' knowing how things should or could be for someone the designer usually does not know. In acts of sympoiesis, i.e. becoming-with, we can apply these skills to engage with those who are not human in our world, to create dialogues and interactions forging human–nonhuman relationships with the power to challenge the current destructive narrative of human superiority (Haraway, 2016). Sympoietic making negates the dominant hylomorphic model of creation ingrained into western thought, where (morphe) is imposed onto inert matter (hyle) to give things — whether material or immaterial — their shape (Ingold, 2009). Creation, in other words, is seen an act of subduing the world with brute force into the desired state.

In contrast, sympoietic making is a gentle probing movement continuously attentive to the agencies and preferences inherent to the world of matter and the other-than human beings populating it in the design process. In the described project, we had to let go of our preconceived ideas, listen and be willing to adapt to things we could not always completely comprehend. We had to develop a subtle relational empathy to enable us to move beyond our own humanness, whilst remaining appreciative of the 'plantness' of our nonhuman other. Through listening to the behaviours of the participating plants we learned to create and design sympoietically, i.e. *with* and not *for* our nonhuman partners.

REFERENCES

- Andersen, L.B., Danholt, P., Halskov, K., Hansen, N.B. & Lauritsen, P. (2015). Participation as a Matter of Concern in Participatory Design. *CoDesign: International Journal of Cocreation in Design and the Arts, 11*(3–4), 250–261. DOI: 10.1080/15710882.2015.1081246
- Bossen, C., Dindler, C. & Iversen, O.S. (2010). User Gains and PD Aims: Assessment from a Participatory Design Project. *Proceedings of the 11th Biennial Participatory Design Conference*. ACM. DOI: 10.1145/1900441.1900461
- Chamovitz, D. (2012). *What a Plant Knows: A Field Guide to the Senses*. New York: Scientific American.
- Grudin, J. & Pruitt, J. (2002). Personas, Participatory Design and Product Development: An Infrastructure for Engagement. *Proceedings of the Participatory Design Conference (PDC2002)*, Sweden. pp. 144–161.
- Halskov, K. & Hansen, N.B. (2015). The Diversity of Participatory Design Research Practice at PDC 2002–2012. *International Journal of Human-Computer Studies, 74*, 81–92. DOI:10.1016/j.ijhcs.2014.09.003
- Haraway, D.J. (2016). *Staying with the Trouble: Making Kin in the Chthulucene*. Durham, NC: Duke University Press.
- Ingold, T. (2009). The Textility of Making. *Cambridge Journal of Economics*, 34(1), 91–102. DOI: 10.1093/cje/bep042
- Karban, R. (2008). Plant Behaviour and Communication. *Ecology Letters, 11*(7), 727–739. DOI: https://doi.org/10.1111/j.1461-0248.2008.01183.x
- Mambrey, P., Mark, G. & Pankoke-Babatz, U. (1998). User Advocacy in Participatory Design: Designers' Experiences with a New Communication Cannel. *Computer Supported Cooperative Work (CSCW)*, 7(3–4), 291–313. DOI: https://doi.org/10.1023/A:1008687122083A

Anette Lundebye

To Name is to Value

He was wearing a polar bear costume made of white fake fur. On his face was an expression of discomfort, like a bad itch you cannot get rid of. The boy's enigmatic expression in the photo haunted me for a while. They say when you dress up as something or someone, you capture the essence of it. The polar bear, the boy's expression, the uncomfortable itch — made it seem as though the boy embodied the itch of extinction. I drew a rendering of the boy as polar bear, which became the starting point for projects exploring biodiversity loss and future generations. The series of drawings titled: 'Silent Stakeholders' was a way to meditate on the miracle of life, but also grieve the disappearance of species.

Silent Stakeholder 1 (2015), graphite on paper, by Anette Lundebye

I first came across the concept of 'silent stakeholders' when teaching 'design for sustainability' to students. Stakeholders are defined as those who are impacted or influenced directly or indirectly by an activity (often commercial). 'Silent stakeholders' are not able to represent their interests in human decision-making, for example: an eco system, a non human species, or unborn humans who do not have a voice (yet), yet who will bear the consequences of current human activities.

Every living system is said to be in decline due to human activity. A common understanding of thinkers who have focused on the interrelationship between humans and the natural world, such as Arne Næss, Joanna Macy and Felix Guattari amongst many others, is how human subjectivity, the environment, and social relations are intimately interconnected. The pain we may feel from our conditions of living are likely to be connected with what is happening in the world we live in. The recognition of what is being lost needs to be expressed in ways that transcend the rational.

The polar bear has become an icon of climate change. As a top predator and a keystone species it plays a unique and crucial role in the maintenance of the Arctic eco-system; its disappearance will cause a domino effect of the disappearance of other species too, decreasing diversity dramatically. We are experiencing a serious biodiversity crisis, scientists say we have entered the sixth major extinction period where species are disappearing. Globally they estimate that species extinction rates have increased over 1,000 times the natural rate (Natural England, 2010: 6). These unnaturally high rates of species loss are a signal — 'the canary in the mine' — of the unsustainable exploitation of our natural environment. Just in my lifetime the Earth's biodiversity will be halved, and biologist E.O. Wilson predicts that we must expect the "future of life" to become globalised and simplified (Wilson, 2002). We have not yet identified all the existing species that co-exist with us on our planet. To date, only between 1.4 and 1.8 million species have been scientifically identified and named (WWF, 2017). Another 8.7 million species are estimated to exist.

They say to name is to value. Carl Linnaeus (1707–1778) was a Swedish naturalist (botanist, physician and zoologist). He is often referred to as the father of taxonomy. He designed a system for naming, ranking and classifying organisms that is still in wide use today. In 1735 he published *Systema Naturae*, the first system of classification. Today this is known as 'biological nomenclature' — a tool to identify and name living things, plants and animals, by giving a scientific name in Latin composed of two parts: the genus and the species. This naming is a way to acknowledge the existence of a species and differentiate it from others, shifting its foreignness to familiarity, giving it a belonging to a 'family'.

In 2016 the artist collective Foreign Investment,[1] which I am a member of, were invited to design an intervention for the launch of an art book titled: *Wild New Territories: Portraits of the Urban and the Wild* (Den Daas & Kenny, 2017) at the Linnaean Society in London. This gave me the opportunity to visit the library to research and to see the first edition of *Systema Naturae* (Linnaeus, 1735), as well as the first English translation of it: *A Genuine And Universal System Of Natural History* (Gmelin, 1816) with illustrations. These books were published in an era of exploration and discovery, at a time when there was abundance of species. Fossils hinted to extinction but this was well before Darwin and the theory of evolution. Although already in 1844, the last couple of Great Auks (*Pinguinus impennis*) became globally extinct. They were killed on Eldey (off the coast of Iceland) to be stuffed and later displayed in the collection of a museum.

The concept for the 'species cards' emerged from the enormous task of identifying and classifying species. We wanted to manifest the loss of biodiversity within the UK. Over the last 200 years, 492 documented species have become extinct and on average 26% of England's species are endangered (Natural England, 2010: 3). As guests would have to sign in upon arrival, we thought this was an opportunity to design a conversation piece that linked the legacy of the past with the present status of some species.

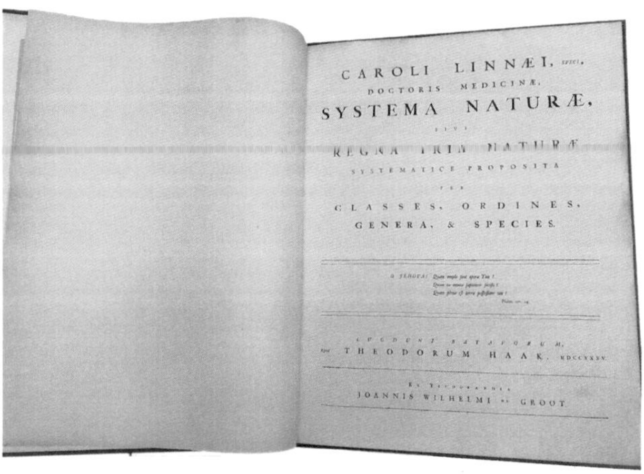

Title page of Systema Naturae (1735) by Carl Linnaeus with permission of The Linnean Society of London and Design template for species card by Foreign Investment.

We designed a set of 100 species cards that each visitor could choose from. Each card had the classifications based on the Linnaean system. The cards were the standard index card size (3 by 5 inches), used for cataloguing. The font used (Adobe Caslon Pro) is similar to the type set used for *Systema Naturae*. In addition three more entries were added:

Habitat: a description of typical places that a species live/d in such as old trees in woodlands, or dwarf shrub in heathland.

Status in the UK: to be stamped either with extinct or critically endangered.

Reason: e.g. Habitat loss, hunting, pesticides.

Guest wearing species card at the Linnaean Society, London, 29 September 2016.

At the top of the card each species (where known) was named by their colloquial 'English' name, then their Latin name. The data about each species were written by hand on the cards, as homage to Linnaeus's daily carefully handwritten notes on paper slips the size of index cards to capture information about plants and animals.

A selection of 50 extinct and 50 critically endangered species in the UK was made according to their current status. We focused on fauna (animals and insects). The research came mainly from Natural England—the UK government's adviser for the natural environment, and the International Union for Conservation of Nature (IUCN), the global authority on the status of the natural world. When compiling the data for the index, and noting the reasons for the classification of either *extinction* or *critical endangerment*, it was striking to find that over 90% of reasons are due to habitat loss. Since the Industrial Revolution many technological advances have led to the significant increase in the rate and scale of environmental change. Destruction and inappropriate management of natural habitats being typical causes. But also persecution of species considered a nuisance or introduction of alien species play a role. For instance, settlements in suburbia are hostile to many species due to pesticides to keep lawns green and appealing fitting with cultural preferences. Consequently, many species have been unable to adapt and many of these species losses and declines were, and still are avoidable.

Having chosen a card with a species, visitors would present it at the table where a member of Foreign Investment would greet them, look at their card and consult with the index of the 100 species, and then stamp it according to its current status. As the list was long it would take some time to identify the species, meanwhile a conversation would take place. Questions from the guests would lead to sighs of dismay. John, one of the many visitors was surprised that the House Sparrow (*Passer domesticus*), a once familiar urban dweller, has completely disappeared from London. Why?

For the rest of the evening each guest would mingle wearing the species card around their neck as a conversation piece to dialogue with others. Perhaps wearing the name of a species might be a step towards more familiarity, making a personal connection? Perhaps the species name transferred some of its essence? Naming is a special form of paying attention, a sign of respect. Perhaps also, representing the species for an evening was a way to remember the whisper of its name when it is gone? Perhaps this recognition gives us a different way of seeing life that is being lost? The polar bear, our children and the millions of species—named and unnamed—are today's silent stakeholders. They have intrinsic value yet no voice. Can design inform the choices that will determine which species will remain and thrive in our future?

NOTES

1 Foreign Investment is an art collective based and founded in the UK in 1996 that develops art projects around notions of exchange, investigating new forms of collaboration in the creation of temporary artworks that are time-based and site-specific.

REFERENCES

— Den Daas, R. & Kenny, K. (2017). *Wild New Territories: Portraits of the Urban and the Wild*. London: Black Dog Publishing Limited.

— Foreign Investment (2016). *LOST, London* [online]. Available at: http://www.foreign-investments.com/main/lost-london-2016/ [accessed 18 March 2018].

— Gmelin, J.F. (1816). *A Genuine And Universal System Of Natural History: Comprising The Three Kingdoms of Animals, Vegetables and Minerals, Arranged Under Their Respective Classes, Orders, Genera, and Species / by the late Sir Charles Linnaeus*. London: Printed for the Proprietor by Lewis & Co.

— IUCN Red List of Threatened Species (2016). International Union for Conservation of Nature [online]. Available at: http://www.iucnredlist.org/ [accessed 15 September 2016].

— Linnaeus, C. (1735). *Systema Naturae, Sive Regna Tria Naturæ Systematice Proposita Per Classes, Ordines, Genera, & Species* (1st ed.). Leiden: Lugduni Batavorum (Haak).

— Natural England (2010). *Lost Life; England's Lost and Threatened Species (NE233)* [online] Available at: http://publications.naturalengland.org.uk/publication/32023 [accessed 24 January 2018].

— Wilson, E.O. (2002). *The Future of Life*. London: Abacus.

— World Wildlife Fund (2017). *How Many Species Are We Losing?* [online]. Available at: http://wwf.panda.org/about_our_earth/biodiversity/biodiversity [accessed 18 November 2017].

(22 February 2017), "My friend!" Crow cried out. "Wait! What can we give you?". "You can give me nothing, Crow. I feel your gratitude. That is enough". Grizzly Bear paused. Then he said, "Be kind".

Louise St. Pierre
DESIGN AND NATURE: A HISTORY

This chapter provides a historical context to ideas and practices of design and nature, highlighting underlying tensions and problematic conceptions about nature in the Modern West. This history differs from broader theories of sustainable design in that I focus specifically on design's relationship with nature. I review various attempts to design with nature in past and recent history, with attention to how design and research are still embedded in a Western conception of our relationship with nature inherited from the Scientific Revolution. Despite aspirations of designers to connect emotionally, philosophically and functionally with the natural world, nature remains subjugated: an 'other'. I argue that with each 'new' approach to designing with nature from the Romantic Movement in the late 19th century, through to contemporary design and current design theory, designers inadvertently continue Modernist and colonialist power relationships that place humans at the top of a hierarchy, with nature at the bottom. These conditions are beginning to change as designers explore ecological theory beyond mainstream influences and as they engage with embodied research in direct relationship with nature.

Design and nature: terms

Contemporary design is a systematic activity that produces artefacts, systems, interactions and communication materials. Anyone who shapes matter, processes and energy to meet perceived needs can be understood as a designer. This text addresses those designers who shape the world through mass-produced culture, and negotiate the intersections between self, society and the natural environment. Design, carried forward through information, media, technology, goods and services, has tremendous impact on the way that human activities impact nature, and on how people perceive nature.

I adopt the definition of nature as the "self-originating material/spiritual world, of which we are a part, including the powers that sustain and govern it" (Bonnett, 2002). Academics and designers dispute the validity of the word nature, as well as the qualities of nature: is it Tainted? Pure? Wild? Powerless? Mute? Artificial? These disputes rarely acknowledge that engaging with nature requires different epistemologies and ontologies than those which were established through the Scientific

Revolution and Enlightenment. A core ambition of this book is to introduce some of these alternate ways of knowing and being.

PART ONE: EARLY VIEWS ON NATURE AND ECOLOGY

Scientific Revolution

Western designers are educated within a worldview that began centuries ago, well before the emergence of many specific disciplines like industrial, landscape, fashion, interaction or communication design. A continuum of changes began in Europe during the Scientific Revolution of the 1500s and continues to shape the work of designers in the Global North.

The Scientific Revolution was a slow transition of belief systems: from a world where, from what we can understand, there was magic and mystery in nature, to one where *if* a phenomenon could *not* be seen and measured, *it was not real*. Before the Scientific Revolution, belief systems such as Organicism held nature in intimate relationship with humans (Merchant, 1990). A spiritual and ritualised relationship with nature was embodied. Illustrations from that time showed plants, humans and other creatures, beings among beings in a non-hierarchical interrelated connection with one another. Nature was considered alive and vital, and humans but one aspect of that complex vitality (Merchant, 1990). In the 1500s, vitalism meant "the unity of matter and spirit as a self-active entity, in which the spiritual kernel is considered the real substance and the material 'cover' a mere phenomenon" (Merchant, 1990: 117). Organicism was a unifying philosophy.

By the 16th and 17th centuries, technological and commercial changes had eradicated the Organicist belief systems that once held nature in intimate spiritual relationship with humans. Images began to appear that showed male beings as superior in the world order. After this, images highlighted only the head of man as dominant, symbolising the emerging belief that male intelligence was of the highest value: the beginnings of rational anthropocentrism. Women gradually became devalued. Rational intelligence, thought to be held in the head, became a standard by which to make decisions: "A new concept of the self was as a rational master of the passions housed in a machine-like body" (Merchant, 1990: 214). Philosophers like Bacon and Descartes believed that craft, innovation and invention would extend rational man's mastery of nature. These beliefs continue to drive design to this day: "*Right, wrong. If this, then that. Causality. Consequences. Looking at parts. Finding solutions. Common sense. Reason over emotion!!*" (St. Pierre Journal October, 2016).

The Scientific Revolution, the Enlightenment and the elevation of scientific and rational thinking combined to diminish society's ability to see mystery and enchantment in the natural world. Westerners no longer saw themselves within an unbounded and uncontrollable cosmos, a world that was larger than humans, a magical world of intrinsic value.

Roots of Modernity

This is where Modernity began: replacing earth-centred values with humanistic ones, replacing multiple local deities with singular, powerful and monotheistic ones, and elevating evidence based ways of knowing. Tlostanova and Mignolo (2009) argue that thought itself was colonised by the elevation of reason; a hegemony so implicit that in contemporary times it is difficult to see or question rationalism. With the rise of rationalism, nature, along with emotion, spirit and the female, were increasingly devalued (Merchant, 1990; Walker, 2013). This is what Plumwood calls 'backgrounding' (Plumwood, 2002).

Design, as most of us understand it today, grew out of the mechanistic worldview, and remains identified with technology, mastery, progress and innovation. Mastery, interpreted either as human control and superiority, or as comprehensive skills, is woven through design's relationship with nature. At various times in history, often in the margins and in the shadows, designers and influencers have questioned the Modern worldview. Escaping the dominant paradigm proves difficult. In prevalent contemporary design we continue to see human centrism, rationalism, assumptions of mastery and little or no reference to spirituality. The persistence of these views undermines designers' attempts to address the needs of the natural world.

The Arts and Crafts Movement: against Rationalism

Led by William Morris, a number of artists, designers and philosophers in the mid-late 19th century advocated a return to nature and mysticism. Known as the Arts and Crafts Movement, they rejected mechanical production and the mechanistic thinking of the industrial revolution. Their works were passionate and emotional. Fine handicrafts, wallpapers and upholsteries were decorated with lush flowers "inspired by close study of nature" (Pevsner 2005: 44). Morris' fiction writing envisions a world where humans are integrated with all of nature, but the work produced at the time belies these ideals. Most are decorative pieces at the service of a privileged mankind, featuring the romance and ideal of nature through line, pattern and sensuality. Some were even toxic, knowingly laced with arsenic, like the *Trellis* wallpaper (Fallan & Jørgensen, 2017: 109).

Fallan and Jørgensen characterise these Victorian designer reformers as "infatuated with nature" (2017: 108), implying that love of nature was an unreasoned and temporary passion. This characterisation underlies much of design and academic scholarship with respect to nature: the term 'Romantic' is used as a dismissive label noting emotional whimsy and superficial styling that is considered irrelevant to the hard-nosed 'business' of design.

Morris' call to align with nature had great resonance for designers, but his ideals were diminished by human-centric and commercial applications. Decorative forms derived from nature were severed from their original context: nature remained an 'other' for mankind's pleasure or manipulation. This story foreshadows design's later explorations of ecological design.

Mechanistic understandings of ecology

At the turn of the 20th century ecological thinking was thought to be a new way of looking at the world. The term ecology was first coined by Haeckel in the late 1800s as "the totality of relations of organisms and the external world" (via Hayward, 1995: 26). Ecological thinking is a potentially holistic or Organicist way of understanding the world, yet definitions are often contradictory. The mechanistic and mathematical epistemologies of the 20th century framed ecology as a practice of itemising and reducing data "abstracted from the organic context in the form of information bits and then manipulated according to a set of differential equations" (Merchant, 1990: 103). Instead of a self-organising and complex set of flows, ecology was seen as a way of controlling and organising nature or society (Anker, 2010; Hayward, 1995). The myth that ecological 'science' can help us manage or control nature explains an underlying disjuncture between design and nature.

Early design and nature

The 1920s and 1930s were a time of sorting out the influence of nature on design in both Europe and North America. While Bauhaus leaders lived in London they engaged in interdisciplinary conversations with biologists (Anker, 2010). These biologists supported scientific technocracy, viewing ecosystems as machine-like systems that could be controlled and managed by humans (Kallipoliti, 2018). As a result, mastery and mechanism permeate the Bauhaus' early engagement with biologists. Biologist Raoul Francé introduced designer Moholy-Nagy to Biotechnik, the notion that plants could offer solutions to technical problems (Kallipoliti, 2018). Moholy-Nagy thought that emulating nature was

1500 Organicism and Mysticism

Scientific Revolution

1600

1700

Industrial Revolution

1800

Arts and Crafts Movement

1900

Ecology as Mechanism

Bauhaus and Biotechnik

World Planning
Human-centred Design

Ecology as Networks
Biomimicry

2000

Bio Design, Biophilic Design
Decentering the Human
Design and Nature

the essence of functionalism. Historian Peder Anker (2010) describes the commitment of Bauhaus pioneers to an ideal of the integration of humanity, ecology and design, with a strong advocacy for humanism and mechanism. The conception of ecology as a set of mechanical rules united these seemingly contradictory ideals. Bauhaus contemporaries Haldane and Huxley advocated "a new mechanistic and mathematical approach in biology as a key material basis for a successful, orderly planning of human society with a new urban matrix of mathematically inspired Bauhaus architecture" (Anker, 2010: 23).

The Bauhaus alignment with rationalism contrasts the emotional connection to nature embodied by Morris and his followers. The Arts and Crafts Movement produced *explicit visualisations* of nature that were later dismissed as 'romantic fashion'. Bauhaus design expressed the perception of an *underlying functionality* of nature. The resulting order, geometry and control became a prevailing influence in design, establishing the conditions for ecological design as mastery.

PART TWO: ECOLOGICAL DESIGN AS MASTERY

Creative control

Ecological design as mastery grew out of myths that humans can be entirely separate from nature and can control natural systems. Attempts to emulate and master nature peak in design for space colonies and *capsule architecture*, the design of self-sufficient spaces for human survival that attempt to cycle human waste in closed systems (Kallipoliti, 2015). Alternatively, designers attempted to *replicate living process* in what they called *world planning*, "the design of the planet itself as much as the design of an object, building, or territory" often extending to "the synthetic replication of natural systems" (Kallipoliti, 2018: 2).

This fascination with independence from the earth system extends to hippies and survivalists, exemplified in the range of tools for survival in the *Whole Earth Catalogues* (Brand, 1968). "We are as Gods and might as well get used to it" heralded Brand in the first catalogue (1968). This assurance of human mastery underlies early developments of ecological design, and continues in varying degrees today. Ironically, the desire to live "in harmony with Earth's ecosystem became for the majority of ecological designers a question of adopting space technologies, analytical tools, and ways of living" (Anker, 2010: 6).

NASA's 1960 Living Pod capsule project was a landmark experiment to hold four men in a capsule that recycled their body waste into food and water. It failed because unaccounted impacts of airborne waste contaminated the equipment. "In addition to carbon dioxide, contaminants in the closed ecosystem of NASA's Living Pod included minute waste particles like dust, hair, skin debris, tobacco particles, odours and toxic substances from cooking, and other formed organic compounds with unpleasant odours like indole, skatole, amines, volatile oils, phenol, nicotine, spores, viruses, and sacrophytic bacteria that decompose organic matter" (Kallipoliti, 2015: 78). The four men inside the capsule suffered from headaches and nausea, and had to be removed from the experiment prematurely. Later capsule architecture projects improved on this experience through tighter planning and computation but still did not demonstrate man's control over nature (Kallipoliti, 2015).

Pioneer ecological designer Buckminster Fuller believed "that only through advanced systems management could one begin to deal with the daunting environmental complexity of the planet Earth" (Kallipoliti, 2018: 19). He advocated the use of technology to solve the world's ecological problems through the computational field of Cybernetics, the compilation of complex data for systems management. Despite his many holistic philosophies, Fuller's conception of himself as inventor inspired by ecological systems aligns him with those thinkers who imagined that nature could be controlled.

Ongoing human-centric influences

Victor Papanek emerged as a hero to young designers in the 70s and 80s when his legendary critique *Design for the Real World* (1972) exhorted them to stop designing 'needless' consumer products and to put their efforts toward less privileged people who truly needed help. His concerns later expanded to a vision for holistic and spiritually grounded design:

"There must be a greater concern for and deeper understanding of nature" (1995: 48). Despite this, Papanek was primarily a champion for human concerns.

Human-centrism permeates design to this day. Like Papanek, most designers understand human-centred design and social design as ethical practices. When left unquestioned, this human centrism tends to address human needs at the expense of other life forms. Designers are often unaware of this disparity, embedded as we are in a society that does not hold humans in intimate connection with nature and does not value all beings equally.

The Brundtland Commission of 1987 did little to shift this relationship. The premise to "meet the needs of the present generation without compromising the ability of future generations" (*Our Common Future* 1987) speaks to the needs of humans above other species. This also perpetuated the assumption that ecological systems could be managed, a view that does not take human fallibility into account. "The rich complexities of the natural world provide a powerful antidote to hubris, for if there are fundamental limits to our knowledge of ecosystem dynamics, we cannot easily 'optimize' our tree farms, nature reserves, or levels of carbon dioxide emissions" (Van der Ryn & Cowan, 1995: 136). Years later, the Sustainable Development Goals (2015) advance strategies for caring for ecosystems with respect yet still position nature as managed by humans who are at the top of a hierarchy. Nature remains conceived as passive, controlled, mute and powerless. While designers are encouraged to consult the Sustainable Development Goals, alternative or post humanist documents that advocate for all species, such as the Earth Charter (Anon, 2000), remain on the periphery of design awareness.

A resurgence of Organicism

Challenges to the mechanistic view of ecology surfaced late in the 20th century and early 2000s. The term ecological design was coined as "any form of design that minimizes environmentally destructive impacts by integrating itself with living processes" (Van der Ryn & Cowan, 1995: x). Capra (2002) described living processes as an interdependent collection of parts within a fluid unbounded complexity of relationships, a multiplicity of nested and circular systems that interact in unpredictable ways. "Wherever we see life, we see networks" (Capra, 2002: 9). This is a nuanced view of ancient Organicism; ecology as a self-animating system of energy flows dispersing "agency and creativity" (Plumwood, 2009).

Designers at the turn of the century understood easily that "any designed product, space, or environment has an expansive presence in the world… projects and extends the presence of all things relative to larger environmental forces and the nexus of global flows" (Kallipoliti, 2018: 2). As designers grappled with this insight, they tried to reduce environmental impacts by making adjustments to materials and manufacturing processes: a pragmatic and incremental approach. More progressive attempts to engage with ecological flows and *integrate with living processes* were undermined by pervasive rationalism, mastery and human-centrism. It is challenging to shake deeply rooted views of humanity's relationship with nature. Designers are further constrained by rationalism within the profession, and externally by the conditions of Capitalism. Despite this, designers persistently demonstrate the intention to practice in relationship with the natural world through many diverse approaches: eco, green, organic, sustainable, biotechnical, biomimetic, bio this… and bio that.

PART THREE: BIO THIS… AND BIO THAT

Bio design

Bio design, where "biology is both the subject and the medium" (Myers, 2012: 195) picks up from Biotechnik, using living nature in a product, as when a grass covered table converts soil energy to charge small electronics. Myers is direct about how innovation furthers human dominance over nature, quoting a designer as saying "One of my goals is to completely design a new life form" (2012: 195). Bio design mixes the astonishment at the workings of nature with a desire to master her. Bürdek, in his *Design: History, Theory* holds similar sentiments, "The cultivation of human organs will make the human body an object to be designed" (2005: 431). The notion of mastery aligned with creativity is an alluring force in design with nature, eclipsing the energy and self-organising capacity of nature herself.

Cradle to Cradle

In the early 2000s McDonough and Braungart also picked up threads of world planning with seductive storytelling in their book *Cradle to Cradle* (2002). Drawing on principles from the field of Industrial Ecology for recycling materials and waste between industries (Graedel et al., 1993), McDonough and Braungart proposed that toxic materials in products could be recaptured and isolated in 'technical nutrient streams' to be used in manufacturing new products. This would theoretically keep toxins from contaminating natural systems. McDonough and Braungart aligned with the business and industry agenda by supporting expansive manufacturing and consumption. They furthered the notion that nature's resources could be harvested for human needs, overlooking the complexities and ecological impacts of the recycling process and the limits of human control.

Human societies unquestionably need to manage their wastes, but this insight becomes distorted when accompanied by advocacy for ongoing unlimited production. Plumwood (2009) characterises plans for containing waste in closed systems as unrealistic "requirements for future human invulnerability and perfection". The participants in the NASA Living Pod capsule followed instructions to the letter and still the capsule became contaminated. Perfection is unattainable. *Cradle to Cradle* is an Arcadian fable.

Biomimicry

Janine Benyus' highly influential book *Biomimicry: Innovation Inspired by Nature* (1997) was met with a powerful and resonant uptake from designers and academics. Echoing the 1930s philosophies of Moholy-Nagy, biomimicry endeavours to model human systems on natural systems, such as wind turbines shaped like a whale fin to reduce energy consumption, with the hope that mechanistic products and systems will then be as sustainable as natural ones. Unlike the technocentric theories of Biotechnik, Benyus included an emotional and spiritual component. Yet as with the theories of William Morris, these ideals became distorted in their application by a focus on solving human problems, and by a deep alignment with economy and industry. This instrumentalises the ideals of biomimicry. Once again nature is being 'used', now as much for the ideas as the materials. Despite having been inspired by nature, the context and framing of Modernism and Capitalism positions most biomimetic artefacts within cultural and industry contexts, leaving them apart from nature.

Biophilic design

The term biophilia was originally proposed in the 1970s by psychologist Eric Fromm as the "love of life" (Andreyev, 2017), although it is widely used to describe a general affinity with nature. Both Fromm and E.O. Wilson, who later popularised the term, emphasise the importance of biophilia to humans (Andreyev, 2017). There are a wide variety of interpretations of biophilia. Biophilic design can be design to enhance people's love of nature, design to inform people about nature or design to offer nature as a service to improve human life.

At its best, biophilic design can draw people out into nature and invite or awaken our innate biophilia, our instinctive affinity with nature. Well designed science apps, for example, help people identify birds, insects and frogs. Through this activity, citizens may become informed, empowered and connected with nature. Beautifully designed nature journals are biophilic. Biophilic design has tremendous potential to open greater awareness and extend the discourse around design and nature. It can inspire an ethics of care.

The realisation of this potential is hampered by human centrism. According to Kellert, biophilic design is practiced primarily "to enhance human well-being" (2018: ix) through the inclusion of natural features such as plants and waterfalls within buildings. This can offer meaningful exposure to nature in urban settings. Kellert lists many benefits for people in contact with nature, including affection, relaxation and creativity. Gillis and Gatersleben make this even clearer with their "A Review of Psychological Literature on the Health and Wellbeing Benefits of Biophilic Design" (2015), which focuses entirely on how biophilic design meets human needs. This reduces biophilic design to the provision of nature as service. Framing nature as an ecosystem service devalues nature and leads to further exploitation of the natural world (Boehnert, 2015).

Learning, practicing and shifting perspectives

Bio design, biomimicry and biophilic design are attempts to see design and nature differently. They represent a search. These different frameworks offer spaces for communities in design and academia to rehearse, critique and learn. I have noted how these learning spaces are constrained by mainstream design practices that prioritize human needs, rationality, mastery, and economic growth.

A number of lesser-known explorations into design and nature have begun to challenge human centrism and design's illusions of mastery. For sake of brevity (and because this book is full of examples) I will mention only a few of them here. These incipient practices are influenced by philosophies that contextualise humans as interdependent within nature, and nature as vital and alive, much like the Organicist thinking of the 14th century. They begin to take us into a more spiritual and philosophical relationship with nature.

Symbiotic Design (Ruano, 2016) is a proposed amalgam of biophilia, biomimicry and resilience theory, hypothetically supporting the development of a symbiotic or holistic designer. Ruano prioritises spiritual wisdom, contemplative practice and observation of nature in his proposed methodology. Acosta and Romeva (2010) advocate in *Ecospheric Design* that we shift the epicentre of design from humans to the ecosphere, "to decentralize human being as unique actor, and… recognizing other actors (the rest of the species) in the project of well-being in Earth" (2010: 35). The authors have been influenced by the non-anthropocentric Earth Charter (Mosquin & Rowe, 2004), and Gaia theory that the earth is alive and self-regulating (Lovelock, 1987). *Bioinclusive Design* draws from the feminist eco-philosopher Freya Mathews (2008) to propose inclusive design that "accommodates both the human and the nonhuman components of the greater life system" (St. Pierre, 2017). This approach is influenced by eco-feminism (Mathews, 2008; Plumwood, 2002) and neo-animism, the awareness of all parts of the world as alive and conscious (Bai, 2013).

These latter examples illustrate how moving beyond mainstream influences leads to progressive conceptions of the relationship between design and nature. An abundance of scholarly thought demonstrates widespread interest in this question as well. Yet here too, we see confusion and hindrances.

PART FOUR: HINDRANCES AND INSPIRATIONS FOR DESIGN AND NATURE

Scholarly hindrances

The legacies of mastery, rationalism and humanism continue to dominate academia, leading to an emotional and intellectual distance that keeps design at arm's length from intimate learning about the natural world. The search for knowledge through intellectual reasoning is a direct inheritance from the Scientific Revolution. In the context of learning about design and nature, it is a barrier to understanding, for it runs counter to physical engagement, humility and curiosity (Jensen, 2008; Vitek & Jackson, 2010; Walker, 2013). Progressive conceptions of design and nature might or might not be supported by academic theory.

Academic statements do not create change. Simple declarations that 'we are part of nature', true as they are, do little to contest longstanding belief systems that allow humans to feel set apart and above the natural world, special and entitled. A single intellectual statement framing interdependence, or even a moment of epiphany, is not enough to challenge human exceptionalism.

Similarly, the claim that there is 'no such thing as nature', that 'nature is in fact a social construct' conveniently dodges the point that everything is then a social construct. Asian philosophers have taught throughout millennia that we only understand the world through the processing of our minds and bodies (Snyder, 2002; Worthy, 2013). The world and all that we know within it is then a biological, psychological and social construction. Coffee cups and cities, children and African-Americans; these are all social constructs. This is the social imaginary, the way human societies categorise and name the world around us. We cannot dismiss Syrians, women or baseball teams by saying they are social constructs. Nor can we do this with nature.

Traditional wisdom has deep knowledge of interdependence, the wisdom that no entity is separate from any other: sentient and non-sentient beings are interconnected and alive. Design academics variously connect to and become distracted by this insight. A wide body of writing about 'New Materialism' invites contemplation about the aliveness of everything around us (Bennett, 2010; Harman, 2018). For many, this extends to be inclusive of manufactured pens, desks and roller skates. This aspect of New Materialism provides an avenue for deeply anthropocentric Western designers to remain focused on that which is produced by and for humans. Reverence for human-created artefacts does not shift our relationship with earth to make ourselves [and our artefacts] "smaller, less central" (McKibben, 1999: xxiv).

The world has been tainted throughout by human activity: widespread pollution, extinctions, environmental toxicity and climate change (McKibben, 1999; Worthy, 2013). The argument that 'pure' nature no longer exists is offered by some as an invitation for more intervention by designers. Dilnot interprets the "expansion of human artifice in every direction" to mean that "the artificial is a matter of the possible" (2015: 121). Within this worldview, humans can continue to act upon nature as masters. It is a position that frames nature as powerless; she has no capacity, creativity or voice of her own. This trajectory is similar to that of colonisers who through their own intervention reduced the capacities of the colonised, then referred to those reduced capacities as rationale to "justify further exploitation and management" (Blenkinsop et al., 2017).

Scholarly activity can be an obfuscation, distraction and ultimately a hindrance to the conversations that we need to have, and to new ways of knowing. This is changing. Valuable theories including decoloniality, eco-feminism, neo-animism and post-humanism begin to shift academic terrain and suggest appropriately humble positions in relation to nature. Decoloniality is a practice that de-links from Eurocentric and Modernist thought rather than simply working as a scholarly transformation from within the academy (Mignolo, 2007). Eco-feminism reorients us to embodied ways of knowing, liberating subjugated knowledge and enabling a relationship with nature as valued (Fawcett, 2000). Neo-animism validates traditional views that life exists in all sentient and non-sentient beings (Bai, 2013), and post-humanism assumes the interdependent importance of all creatures (Haraway, 2016). The academy is also beginning to admit personal writing, direct lived experience, intimacy and vulnerability into the discourse.

Practising humility

On a recent walk along the shores of False Creek in Vancouver led by indigenous guide Nicole Preissl of the Sto:lo First Nation, a group of our students were taught to make rope from the autumn remains of iris leaves (Preissl, 2018). As Preissl told stories of the land, students twined the strands with their hands, creating unexpectedly beautiful fine ropes and bracelets. At the end of the walk, we learned that we could not keep these treasures. The Sto:lo Nation, who lived in harmony with nature for ten thousand years believe that we as individuals do not own artefacts. When one acquires artefacts or knowledge, one is considered rich; this wealth confers the responsibility to give to others (Preissl, 2018). The ropes were gifted to our guide, who promised they would be woven into an installation that would biodegrade over the winter. This is a practice of humility and relation. Weaving leaves, walking, talking and giving back are learning practices done in a supportive community, reinforced over time. It is a ritualised and embodied learning, not an intellectual one.

Emerging practices that may support this type of learning include slow design (Pais & Strauss, 2016) and practices of care (Puig de la Bellacasa, 2017). These practices are specific and place-based. Design for ecological relationality is deeply context dependent; valid and proper knowledge comes from the context in which it occurs. Only with deep understanding of specific local conditions can we begin the relational work of tending (Berry, 2002). Van Der Ryn and Cowan (1995) recount a story of a Chilean farmer's wife who noticed that the butter was a bit sour and from that taste knew that the problem lay in the cow's diet. She was able to recommend to her husband appropriate crops to correct the field's nutrient imbalance. This may seem an aspirational degree of connection to nature's self-maintaining rhythms, yet indigenous communities lived this Organicist and deeply spiritual connection with the land for centuries.

Design, nature and spirituality

Stuart Walker asserts that design needs to reclaim a wide range of spiritual practices that are "holistic in that they can affect all aspects of life" (2017: 97). The spiritual view of design and nature involves attending to the kind of connection with nature that emerges from the awareness of ourselves as but "one manifestation of a web of relationships which encompasses everything" (Loy via Bai, 2001). Plumwood (2009) describes this as the self-organising creativity of ecological systems. For Plumwood, this was inevitably spiritual, encompassing the many diverse life forms. She proposes that spirituality can be considered "in dialogical terms, as a certain kind of communicative capacity that recognizes the elements that support our lives" (2002: 220). This suggests a full engagement with and respect for the natural world. Blenkinsop articulates the value of turning to nature as a teacher: "New relationships and commensurate language will arise slowly out of action—actual engagement in new ways of being present to, and interacting with, the world" (Jickling et al., 2018: 36).

Contemplative traditions like Buddhism also practise a dialogical relationship with all beings, animate or inanimate. Indigenous practices know the world as full of vibrant spirits. These spirits can take the forms of mountains, trees, ravens, coyotes: beings that are related to us and intertwined with us. This is what Plumwood means by multiple centrings instead of human-centring, a rich and varied complexity of life to be honoured (Plumwood, 2002). Spirituality is in the connections, the organisation of this complexity. Honoring all of this brings us into what Buddhists call *right relationship* with Earth (Nhat Hanh, 2013).

Concluding thoughts

This chapter has described how designers throughout history have been searching for ways to design with nature. I have discussed how designers' attempts to work with nature have been limited by worldviews and power relations that were established centuries ago. These ontologies are being challenged by theories and philosophies that shift implicit assumptions about power and mastery. This can open designers to humility, understanding and relationship. In this book we offer examples of engagement with nature that move these theories into being; *moments* of action, of open and humble exploration between design and nature. *Moments* that begin to enact new terms of engagement for design and nature.

NOTES

1 Sustainable Design has many definitions, most of which stem from the Brundtland Report of 1987: "meeting the needs of the present without compromising the ability of future generations to meet their own needs", a very human-centric and progress-oriented definition.

REFERENCES

— Acosta, G.G. & Romeva, C.R. (2010). From Anthropocentric Design to Eco-spheric Design: Questioning the Epicentre. Presented at the International Design Conference-Design 2010, Dubrovnik, Croatia.

— Andreyev, J. (2017). *Biophilic Ethics and Creativity with More-Than-Human Beings*. PhD Thesis, Simon Fraser University, Burnaby, Canada. Retrieved from http://summit.sfu.ca/identifier/etd10108

— Anker, P. (2010). *From Bauhaus to Ecohouse: A History of Ecological Design*. Baton Rouge, LA: Louisiana State University Press.

— Bai, H. (2001). Challenge for Education: Learning to Value the World Intrinsically. *Encounter, 14*(1).

— Bai, H. (2013). Peace with the Earth: Animism and Contemplative Ways. *Cultural Studies of Science Education, 2*(8). https://doi.org/10.1007/s11422-013-9501-z

— Bennett, J. (2010). *Vibrant Matter: A Political Ecology of Things*. Durham, NC: Duke University Press Books.

— Benyus, J. (1997). *Biomimicry: Innovation Inspired by Nature* (1st ed.). New York: William Morrow and Co.

— Berry, W. (2002). Conservation is Good Work. In T. Butler (Ed.), *Wild Earth* (pp. 142–155). Minneapolis, MN: Milkweed Editions.

— Blenkinsop, S., Affifi, R., Piersol, L. & De Danann Sitka-Sage, M. (2017). Shut-Up and Listen: Implications and Possibilities of Albert Memmi's Characteristics of Colonization upon the Natural World. *Studies in Philosophy and Education*, 36(3), 349–365.

— Boehnert, J. (2015). Cutting up the Commons: The Violence of Epistemological Error. *Journal of Visual Culture, 14*(2), 200–204. https://doi.org/10.1177/1470412915594337

— Bonnett, M. (2002). Education for Sustainability as a Frame of Mind. *Environmental Education Research, 12*(3–4), 265–276. https://doi.org/10.1080/13504620600942683

— Brand, S. (1968). The Purpose of The Whole Earth Catalogue. Retrieved 25 January 2019, from http://www.wholeearth.com/issue/1010/article/196/the.purpose.of.the.whole.earth.catalog

— Bürdek, B.E. (2005). *Design: The History, Theory and Practice of Product Design by Bernhard E. Burdek*. Basel-Boston-Berlin: Birkhäser Architecture.

— Capra, F. (2002). *The Hidden Connections: Integrating the Biological, Cognitive, and Social Dimensions of Life into a Science of Sustainability* (1st ed.). New York: Doubleday.

— Dilnot, C. (2015). The Matter of Design. *Design Philosophy Papers, 13*(2), 115–123. https://doi.org/10.1080/14487136.2015.1133137

— Fallan, K. & Jørgensen, A. (2017). Environmental Histories of Design: Towards a New Research Agenda. *Journal of Design History*, 30(2), 103–121.

— Fawcett, L. (2000). Ethical Imagining: Ecofeminist Possibilities and Environmental Learning. *Canadian Journal of Environmental Education* (CJEE), 5(1), 134–149.

— Gillis, K. & Gatersleben, B. (2015). A Review of Psychological Literature on the Health and Wellbeing Benefits of Biophilic Design. *Buildings*, 5(3), 948–963. https://doi.org/10.3390/buildings5030948

— Graedel, T.E., Allenby, B.R. & Linhart, P.B. (1993). Implementing industrial ecology. *Technology and Society Magazine, IEEE, 12*(1), 18–26. https://doi.org/10.1109/44.192717

— Haraway, D.J. (2016). *Staying with the Trouble: Making Kin in the Chthulucene*. Durham, NC: Duke University Press Books.

— Harman, G. (2018). *Object-Oriented Ontology: A New Theory of Everything*. London: Penguin UK.

— Hayward, T. (1995). *Ecological Thought: An Introduction*. Cambridge, UK: Polity Press.

— Jensen, J. (2008). Educating for Ignorance. In B. Vitek & W. Jackson (Eds.), *The Virtues of Ignorance: Complexity, Sustainability, and the Limits of Knowledge* (pp. 293–306). Lexington, KY: University Press of Kentucky.

— Jickling, B., Blenkinsop, S., Timmerman, N. & Sitka-Sage, M.D.D. (Eds.). (2018). *Wild Pedagogies: Touchstones for Re-Negotiating Education and the Environment in the Anthropocene* (1st ed.). Switzerland: Palgrave Macmillan.

— Kallipoliti, L. (2015). Closed Worlds: The Rise and Fall of Dirty Physiology. *Architectural Theory Review, 20*(1), 67–90. https://doi.org/10.1080/13264826.2015.1078385

— Kallipoliti, L. (2018). History of Ecological Design. *Oxford Research Encyclopedia of Environmental Science*. https://doi.org/10.1093/acrefore/9780199389414.013.144

— Kellert, S.R. (2018). *Nature by Design: The Practice of Biophilic Design*. New Haven, CT: Yale University Press.

— Lovelock, J. (1987). *Gaia: A New Look at Life on Earth*. Oxford, UK; New York: Oxford University Press.

— Mathews, F. (2008). Thinking from within the Calyx of Nature. *Environmental Values, 17*(1), 41–65.

— McDonough, W. & Braungart, M. (2002). *Cradle to Cradle: Remaking the Way We Make Things* (1st ed.). New York: North Point Press.

— McKibben, B. (1999). *The End of Nature* (2nd ed.). New York: Anchor Books.

— Merchant, C. (1990). *The Death of Nature: Women, Ecology, and the Scientific Revolution* (2nd ed.). New York: HarperOne.

— Mignolo, W.D. (2007). Delinking. *Cultural Studies, 21*(2–3), 449–514. https://doi.org/10.1080/09502380601162647

— Mosquin, T. & Rowe, S. (2004). A Manifesto for Earth. *Biodiversity*, 5(1), 3–9. https://doi.org/10.1080/14888386.2004.9712713

— Myers, W. (2012). *Bio Design: Nature. Science. Creativity*. London: Thames & Hudson.

— Nhat Hanh, T. (2013). *Love Letter to the Earth*. Berkeley, CA: Parallax Press.

— *Our Common Future* (1987). Retrieved January 23, 2019, from http://www.un-documents.net/wced-ocf.htm

— Pais, A.P. & Strauss, C. (Eds.). (2016). *Slow Reader: A Resource for Design Thinking and Practice*. Amsterdam: Valiz.

— Papanek, V. (1972). *Design for the Real World: Human Ecology and Social Change*. New York: Thames & Hudson.

— Papanek, V. (1995). *The Green Imperative: Natural Design for the Real World*. New York: Thames & Hudson.

— Pevsner, N. (2005). *Pioneers of Modern Design: From William Morris to Walter Gropius*. Bath, UK: Yale University Press.

— Plumwood, V. (2002). *Environmental Culture: The Ecological Crisis of Reason* (1st ed.). London; New York: Routledge.

— Plumwood, V. (2009). Nature in the Active Voice. *Australian Humanities Review*, issue 46. Retrieved from http://australianhumanitiesreview.org/2009/05/01/nature-in-the-active-voice/

— Preissl, N. (2018, November). *False Creek Fibershed Walks*. Presented at the INDD 200, Vancouver.

— Puig de la Bellacasa, M. (2017). *Matters of Care: Speculative Ethics in More than Human Worlds* (3rd ed.). Minneapolis, MN: University of Minnesota Press.

- Ruano, D.S. (2016). *Symbiotic Design Practice: Designing With-in Nature*. Doctoral Thesis, University of Dundee, Scotland. Retrieved from http://discovery.dundee.ac.uk/portal/en/theses/symbiotic-design-practice%284c3d61c4-5524-45be-8662-8fdabe2517b0%29.html
- Snyder, G. (2002). Is Nature Real? In T. Butler (Ed.), *Wild Earth: Wild Ideas for a World Out of Balance* (pp. 195–198). Minneapolis, MN: Milkweed Editions.
- St. Pierre, L. (2017). Bioinclusive Pedagogy: Redo Our Relationship with Nature. *Cumulus Kolding Proceedings*. Retrieved from http://cumuluskolding2017.org/proceedings/
- Sustainable Development Goals (2015). Retrieved 20 January 2019, from https://sustainabledevelopment.un-.org/?menu=1300
- *The Earth Charter* (n.d.). Retrieved 23 January 2019, from http://earthcharter.org/discover/the-earth-charter/
- Tlostanova, M., and Mignolo, W. (2009). Global Coloniality and the Decolonial Option. In *Kult 6, Special Issue: Epistemologies of Transformation* (Fall) Department of Culture and Identity, Roskilde University.
- Van der Ryn, S. & Cowan, S. (1995). *Ecological Design* (2nd ed.). Washington, DC: Island Press.
- Vitek, B. & Jackson, D. W. (Eds.). (2010). *The Virtues of Ignorance: Complexity, Sustainability, and the Limits of Knowledge*. Lexington, KY: University Press of Kentucky.
- Walker, S. (2013). Design and Spirituality: Material Culture for a Wisdom Economy. *Design Issues*, *29*(3), 89–107.
- Walker, S. (2017). *Design for Life: Creating Meaning in a Distracted World*. Abingdon, UK; New York, NY: Routledge.
- Worthy, K. (2013). *Invisible Nature: Healing the Destructive Divide Between People and the Environment*. Amherst, NY: Prometheus Books.

SECTION III: STANDING

Standing to achieve a view

In standing, humans rise up from earth to achieve an uplifted, reflective view of it. Standing like a tree supports natural wakefulness and effortless attention. When we are standing, we are not protected as when sitting, sheltered. We may sway with strong wind, feel the cold, feel discomfort. The five contributions in *Standing* bring examples of both the benefit of overview that standing affords and the experience of friction. Here we may face shameful pleasure of supremacy over nature, shameful fear of life itself. The role of us upright animals—humans—in the deterioration of vital earth systems is made explicit in references to capitalism, speciesism, fossil fuel dependency.

In the first essay in this section, Markus Wernli sees a urine explosion as a happy accident that took an "overconfident" design project aimed at growing lettuce indoors with residents' own, self-processed urine in Hong Kong, to a fruitful space of vulner-able design. When designers and participants could relinquish control of nature, it was possible to embrace design "not as practice to shield us from the unpredictability of living but to engage and grow with it".

Katarina Dimitrijevic presents stark facts about how the petroleum based society has given birth to a new hybrid environment, the plastisphere, oceanic living landfills with plastic particles integrated into the marine food chain. Katarina engages design stakeholders in the UK in exploring these sinister realities of human–culture–nature entanglements, and in seeking perspectives of other species to find pathways out of petrol and plastic dependency.

Else Skjold and Frederik Larsen position fur as a system of manipulation, in an essay that sometimes shockingly demonstrate man's dominion over nature. Else and Frederik bring to the forefront how aesthetic ideals and processes of industrialisation literally have redesigned nature. They also point to the potential of fur as a system of care and a learning space, which can ask core and situated questions about the value of life, nature and craft.

In his designerly pursuits to explore "less violent relations with fellow critters", Erik Sandelin, Sweden, introduces us to the character of Bully. Erik proposes that Bully can help us in essential engagement with uncomfortable sides of human–nature interactions, such as the enjoyment some humans take in their supremacy, exemplified with the mundane act of fishing.

This section ends in a forest in the UK, with Mathilda Tham's critique of design as a war against ontological, epistemological as well as ecological diversity. She offers a celebration of a potential life affirming design, that relinquishes control, heroics, predefined outcomes. She proposes that a key question for life affirming designers should be 'how did you keep yourself dirty today?'

(22 March 2017) STILL raining here. We are beginning to dissolve.

Markus Wernli

Short-comings and Vulner-abilities

In a somewhat speculative, participatory study we tried to design a consequential entanglement between people and plants. "We" refers to a PhD student (the *fermentation enthusiast*) and a product designer (the *horticulture enthusiast*) and 22 plant-loving participants. Over two months in spring 2017 we practically explored our personal role in bio-material circulations. Yet our negligence for the basic needs of nutrients-transforming bacteria made us extremely susceptible to floundering plants and inconvenient confessions, and postponed little breakthroughs.

Intentional consequences

We wanted to see what happens when residents in Hong Kong grow lettuce indoors simply from their own, self-processed urine. Yes, our urine. Some voices—that we respect—told us not to implicate people in their own sanitation. We didn't listen and prototyped a tiny niche for our nutrients-abundant outflow (Wolgast, 1993), crafting it into the life force it is biologically meant to be. We wanted to explore *life-serving alternatives* (Hayes, 2010) in contrast to narrow 'waste treatment' options where sludge is mined for phosphorous, biomass is gas-harnessed into abstract energy and runoff is detrimental. We wanted to temporarily reclaim our 'agro-ecological' selves (DuPuis, 2015) we used to be prior to pre-urbanism.

We observed how people bring living plants into homes and raise them, for decoration or for food supply. Many Hong Kong residents live in crammed, dark flats, so recently these *hydroponic* plant incubators became all the rage: Plants are bred in a petrochemical, aerated nutrients solution and with LED lighting—in just half the time of soil-based growing (Trejo-Téllez et al., 2012). Soil and sunlight are no longer needed.

We also noticed how some people are secretly adding a splash of their fresh urine to the water in the vase or pot of tomatoes as a frugal way of fertilising. Such life-hack practices produce rather mixed results. The volatile urea reacts with the air, causing malodours instead of mineralising into plant nourishment (Jones, 1997: 82).

Engaging urine and people

We went about combining *urine prudence* with *hydroponic magic* and dared to shrink-wrap the food loop for domestic use. We gathered expertise from microbiologists. We consulted with health care specialists. We had wet horticultural dreams with ecological engineers. We perfected urine-to-hydroponics upcycling in year-long experimentations. And we gave the thing a name: ANTHROPONIX. "ANTHRO" paid tribute to materials made and discharged by humans; "PONIX" referred to the hard labour (toil) that growing food entails. We hoped to *con*-template—observe *together*—how lifestyles and substances penetrating our bodies impact flourishing and taste of the vegetal offspring as indicated in figure 1.

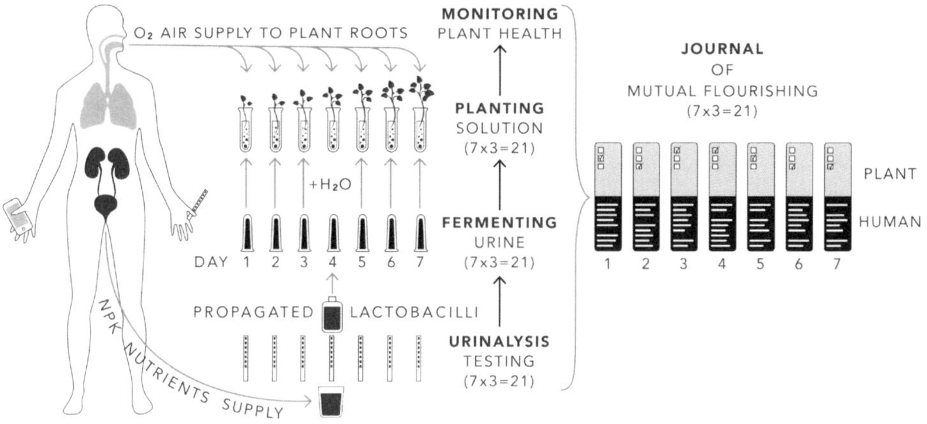

Figure 1: Tiny urine specimens are Urinalysis-tested, then fermented and diluted in water to grow lettuce while the welfare of human and plant is recorded in a detailed journal over a six-week period.

Adopting DIY lactic acid fermentation, akin to yoghurt cultivation, we were able to stabilise our urine (Andreev et al., 2017). Using bacterial cultures, in-vessel storage and three weeks of fermentation is ridding urine of its smell, making it fit for indoor use. We could not afford costly bio-lab testing, so Urinalyses dye strips let us monitor a few medical parameters. Every day we collected, tested and fermented 20ml of our morning urine at home over three weeks. After fermentation, a lettuce or watercress seed was grown in each diluted specimen. To track how one's eating habits and emotional fluctuations impact the urine-bred plant life, we kept an elaborate journal.

We imagined a horticultural co-design where people, plants and bacteria would happily collaborate toward shared prospering. This urine/plant cultivation was collectively held together by five bi-weekly workshops to provide instalments of supplies, skilling and sharing opportunities. Using a multi-modal aesthetic that juxtaposed scientific with philosophical considerations, the workshops intended to bridge the disconnects between human and plant, design and nature.

Reality comes more agitated

On April fool's day we launched ANTHROPONIX with 22 participants. For the consecutive 21 days, we collected tiny bits of our morning urine — 'midstream' — in centrifugal test tubes. The 'centrifugal' made us believe the tubes would contain the gases released during fermentation when we ordered them the week before. Yet, the carbon dioxide of hard-working lactic acid bacteria was stronger than our centrifugal lids. Most of us were surprised to find urine sample no. 1 dispersed in the bathroom when returning home in the evening. Only hours after the project began, the facilitators abandoned their designer's pride and launched fearlessly into damage control: admitting the lack of preparation, giving explanations, imploring to duct-tape the lids.

Miraculously, the participant count remained at 22 on day two and throughout the venture. The urine blow-up was a *happy accident*. The chemical reaction made the urine transformation palpable: it did *not* smell. Since everybody encountered the same problem, it was our heightened moment of group initiation, a bonding that proved to be vital for bigger challenges still to come.

Our designerly thriftiness and overconfidence led to a second design flaw that would make the plant growing into a speculative scramble. When we down-scaled the planter size (for budgetary reasons) we ignored how nutrients-transforming bacteria could be pushed beyond operational limits. The nutrients in urine need to be first mineralised by oxygen-hungry bacteria before plants can utilise them. To prevent suffocation from nutrients overload (eutrophy) constant air injection was needed which made it challenging to grow lettuce, beans and watercress.

Staying engaged on crisis point, most of us responded in creative ways to nurture and rescue the floundering plants. Often, *not* following the facilitators' directions proved to be most successful. One participant systematically germinated different seed varieties until she succeeded with a particular kind of bean that tolerated the restrictive planters. Others tried to better understand the biochemical processes involved and researched ways for monitoring and improving the nutrients solution. People with their own hydroponic systems began to supplement our organic solution with the petrochemical option in a desperate effort to re-nourish the starving plants. Veteran gardeners in the group funnelled their fruitless attempts into a teachable moment for the facilitators and pointed them to horticultural books and knowledge resources for considering alternative planting methods.

While most of the activities took place at our homes, many referred to the project as a *camp experience* due to its intensity by overcoming uncertainties, knowledge-packed workshops and heated group discussions. These extraordinary circumstances over a limited period of time warranted extraordinary efforts in our busy lives. When our shrink-wrap food loop (figure 2) unravelled, we realised how there was value in becoming vulnerable.

The engaging precarity

Figure 2: More-than-human health interaction through a closed food loop with plants growing on fermented urine.

In a venture so technically limiting, the ANTHRO, as in *humanly caused*, and the PONIX, as in *overcoming obstacles*, took on real world meaning. The open-ended struggle in this socio-vegetal precarity shifted our experience. On the flip side of precarity and distress can be joy, belonging and meaning as Brené Brown (2013) indicates. This vulnerability has enlivened and activated us, the organismic, socio-technological, cultural and emotional beings that we all are (Sonne & Tønnesvang, 2015).

Reusing our urine made us, on a foundational level, vulner-able as *organismic beings*. Environmental toxins and non-degradable polluters such as microplastics in our tap water (Tyree & Morrison, 2017) accumulate uncontrollably as *body burdens* (Liboiron, 2016), inevitably ending up in our urine and plants. We do have only limited control over the intake of foods, drugs and body care products. Pharmaceuticals can linger in our organisms for weeks (Mullen et al., 2015) and antibiotics are sabotaging the fermentation process for months (Sollazzo et al., 2014). One participant thoughtlessly took flu meds during collection period and found himself worrying of being trapped in an eternal pharmaceutical feedback loop if he ate the beans grown in his urine.

Trying to align our eating habits towards plant flourishing and withholding 'lucky tiny urine samples' from spilling into the sewer, made us aware how we are vulnerable as *socio-technological beings*. When weaving ourselves into the food web, we better sensed the dependency on that Australian farmer growing our organic oats and the food system in between. It made us more grateful to the sanitation worker and logistics that enable us to use a toilet. We also realised how our relationships with non-human actors like lactic acid bacteria and lettuce sprouts come without guarantees and are unpredictable. Yet, collaborating with them requires our dedication.

Due to the lack of tangible—edible!—outcomes, people started to worry about the legacy of the project. In this way we became vulnerable as *cultural beings* who have a need for shared sense-making. The concluding interviews became often an opportunity for charting out more radical and regenerative flora–fauna interactions in Hong Kong. Participants expressed a desire to contribute to a larger narrative and felt obliged to the 'greater good'. Many felt like they had a stake in the implications of this project by taking a stand and differing from societal norms.

Witnessing the struggle and floundering of the plants made us discover our vulner-ability as *emotional beings* who try to make sense of the complexity called life. The suffering of the plants was able to trigger existential confessions in some. True for all was how beyond attitudes and inhibitions, we can engage on a deeper level with our urine, our senses, and our little personal journeys on this metabolising planet. It took courage to bring forth this vulner-ability. In the least, the imperfect design of ANTHROPONIX made for rather perfect con-*versations* (*dancing* together).

Stick-with-it 'phase 2'

Through this collective stumbling-forward in uncertainty, some kind of accelerated group learning emerged. To ensure all dying plants, urine concoctions and efforts were not in vain, participants had a central role in designing a more plant-friendly 'phase 2'. In consensus with gardening-savvy group members we scrapped the original concept and aimed at a hybrid planting approach. The water-dissolved human nutrients are now drenching a coco-peat growing medium that is much more conducive to bacterial-cationic activity. It makes possible plant raising without external air supply on different scales and configurations. 'Phase 2' kicked in before the first had ended and still continues to date.

ANTHROPONIX ran contrary to a culture telling us we need to be perfect, predicting and controlling. Coming to grips with vulner-ability, we needed to learn how to accept shortcomings as a mode of life-affirming agency. Together with metabolising humans and nonhumans we had redefined design as vulner-ability practice where we are going-through, crafting-through, waiting-through the (non)outcomes and bear with all the consequences—self-inflicted or not—more wholeheartedly. Rather than distracting ourselves with right–wrong discussions, or, seeking to avoid the messiness with narrow perfectionism, we took a less glorious position of the insufficient, yet possibly more authentic co-designer. We were lucky to find lettuce, urine and fellow participants who gave us the courage to be gracefully imperfect. It made it possible to enrich and complete the experience with each other through dedication, bonding and meaning. It made the outcome so much more fermentive and probiotical.

This essay looks at design not as practice to shield us from the unpredictability of living but to engage and grow with it. Recent research in social psychology demonstrate how first-hand experience of vulnerability is necessary for human flourishing (Sonne & Tønnesvang, 2015). What makes us vulnerable—our human susceptibility to discomfort, pain, failure—is not only a source of struggle, shame and inferiority. It is also the origin of pleasure, creativity, belonging and sense-making. Embracing our vulnerability—rather than eliminating it—is a potentially more integrative way of accessing change and unleashing human development that stems from within ourselves. We have a sense that this article just scratched the surface of what the fuller implications of vulner-ability in design practise might be. Co-designing for and through the fermenting, sprouting, living world has augmented that. In the least, finding mutual support and affirmation when being biosocially exposed may help us transcend the design–nature duality.

Acknowledgements: ANTHROPONIX was made possible with a seed grant from Design Trust in Hong Kong, an internationalisation grant from Dutch Creative Industries NL in Rotterdam and supported by the Research Institute for Sustainable Urban Development at The Hong Kong Polytechnic University.

REFERENCES

— Andreev, N. et al. (2017). Lactic Acid Fermentation of Human Urine to Improve Its Fertilizing Value and Reduce Odour Emissions. *Journal of Environmental Management*, 198, 63–69. DOI: 10.1016/j.jenvman.2017.04.059.

— Brown, B. (2013). *Daring Greatly: How the Courage to Be Vulnerable Transforms the Way We Live, Love, Parent, and Lead*. London: Penguin.

— DuPuis, M. (2015). *Dangerous Digestion: The Politics of American Dietary Advice*. Oakland, CA: University of California Press.

— Hayes, S. (2010). *Radical Homemakers: Reclaiming Domesticity from a Consumer Culture*. Richmondville, NY: Left to Write Press.

— Jones, A. (1997). *Environmental Biology*. New York: Routledge.

— Liboiron, M. (2016). Redefining pollution and action: The matter of plastics. *Journal of Material Culture*, (50) 1. 87-110. Available at: https://journals.sagepub.com/doi/full/10.1177/1359183515622966

— Mullen, R. et al. (2015). Analysis of Pharmaceuticals in Food Crops Grown in Urine and Related Products Fertilized Soil. [online]. Available at: http://richearthinstitute.org/wp-content/uploads/2015/10/Pharmaceuticals-in-Urine-and-Related-Products-ACS-Poster.pdf [accessed 5 March 2018].

— Sollazzo, A. et al. (2014). *Usine Du Trésor Noir: User's Manual*. Brussels: Collective Disaster.

— Sonne, M. & Tønnesvang, J. (2015). *Integrative Gestalt Practice: Transforming Our Ways of Working with People*. London: Karnac.

— Trejo-Téllez, L. et al. (2012). Nutrient Solutions for Hydroponic Systems. In T. Asao (Ed.), *Hydroponics* (pp. 1–23). London: IntechOpen.

— Tyree, C. & Morrison, D. (2017). *Invisibles: The Plastic Inside Us* [online]. Available at: https://orbmedia.org/stories/Invisibles_plastics [accessed 5 March 2018].

— Wolgast, M. (1993). *Rena Vatten: Om Tankar I Kretslopp*. Uppsala: Creanom.

Katarina Dimitrijevic

Living Landfill

Guided by scientific findings, including global waste data, I argue that the oceans are the world's largest mismanaged landfill and unprotected ecosystem on the planet (Boonzaier & Pauly, 2015: 30). This is hard to understand because we normally connect landfills with the land. It is also difficult to reverse and therefore requires new ways of engagement. In this essay, I will share some of my explorations aimed at offering a more-than-human perspective on oceanic plastic pollution. I will also invite readers to visualise the ocean in novel ways—as the living landfill.

We live in a throw-away society, where world cities generate about 1.3 billion tonnes of solid waste per year (Hoornweg & Bhada-Tata, 2012: VII). Approximately 11% of the waste generated is plastic, with 4.8 to 12.7 million metric tonnes entering the ocean (Jambeck et al., 2015: 768–771). Plastics production has surged over the past 50 years, from 15 million tonnes in 1964 to 311 million tonnes in 2014, and is expected to double again over the next 20 years (MacArthur Foundation, 2017: 11).

'Plastisphere' (Zettler et al., 2013: 7137–7146) is a scientific term for a new bio-layer in the oceans, created by human discard of plastics. Currently, the oceans have been estimated to contain 269,000 tonnes of plastic (Eriksen et al., 2014: Results). The term Plastisphere describes the variety of microbes living and thriving on this new oceanic habitat consisting of 5.25 trillion plastic particles.

This essay draws on *Plastic Soup*, one of the participatory design workshops which form part of my PhD work. The purpose of these workshops is to narrate the material and social afterlife of plastic packaging; that is the life after human use, but very much part of the current life in the oceans. The workshops constitute an action based platform that supports plastic waste conversations. These conversations are about raising awareness of plastic pollution and building empathy with nonhuman species who are entangled in plastic waste. Plastic pollution provides the context in which I engage with the UK based Higher Education Design stakeholders. Through the workshops, I am inviting designers to visualise the oceans' plastic pollution and by doing this shifting their point of view to be more inclusive of more than human species.

Plastic Soup

> *We turn to the ocean itself: to its three-dimensional and turbulent materiality, and to encounters with that materiality.*
> (Steinberg & Peters, 2015: 247–248)

The 'Plastic Soup' workshop formed part of the "*PhD By Design*" conference at Goldsmiths, University of London (PhD By Design, 2015: 48–49). The participants consisted of six graduate students from across the UK, five from design research and one from political science. During the workshop, the participants engaged in cooking plastic soup while having waste-centric conversations. The workshop explored the properties of the North Atlantic Gyre and in particular the Barents Sea Patch, which is the newly identified sixth gyre (Van Sebille & Froyland, 2012: 1–6).

There are eleven gyres described in the world's oceans. Gyres are large-scale systems of wind-driven surface currents, caused by the Coriolis effect (rotating system perpendicular to Earth's axis) (Eriksen et al., 2016: 1.1). As part of exploring the ocean space, the participants were asked to visualise two gyres closest to Northern Europe, the North Atlantic Gyre and the subpolar Barents Sea. The workshop paid particular attention to the Barents Sea Patch. While this patch is accumulating, it is still passive, which means that it has yet to start rotating—forming a gyre. The group was asked to visualise this passive oceanic soup, which is not just floating under the surface of the sea, but at all depths: recently, scientists have described documented phenomena akin to plastic *smog*, that is slowly settling to the seafloor (Eriksen et al., 2016: 7.0).

We filled a large plastic tank with lukewarm salted water to simulate the sea environment. I had brought with me a selection of plastic waste. We added used food packaging, plastic bags, balloons, net, rope and plastic animals. We chopped, confetti punched and cut the plastic into small pieces. All was added to the water to simulate three processes of degradation at play in the oceans: a) mechanical, b) chemical and c) biological (Eriksen et al., 2016: 1.1). The chopping and punching sped up the degradation and fragmentation of the plastic, which would happen over time in the gyres. Next, one of the participants stirred the plastic soup mix with her hand to mimic vertical and horizontal oceanic currents. Almost instantly, the whole lot had entangled. Entangled, it was difficult to break individual elements free. It made us think of a sea turtle caught in a fishing net (fig.1., top view). The colour and the rotational movement of the plastic was beautiful and lethal at the same time. The hand-induced current stimulated the rapid drop of the plastic to the bottom of the tank. Sedimentation is the ultimate fate for plastic lost at sea (Eriksen et al., 2016: 3.2). The rotation created a cloud of buoyant plastics gathered under the top of the water surface. The plastic tank improvisation

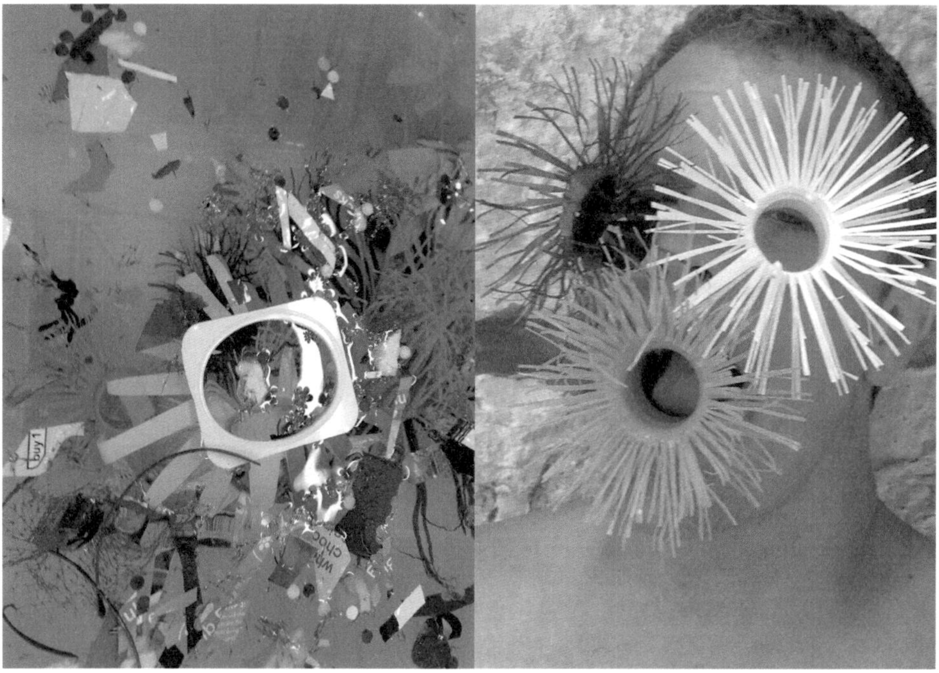

Figure 1: Plastic Soup: Barents Sea Patch, workshop at PhD By Design (KraalD, 2015), Goldsmiths, University of London, UK. Left: top view; Right: side view. Selected waste media: reused mixed plastics, plastic toys, tap water and sea salt.

reflected aspects of the dynamics within the gyre, confirmed some scientific facts, such as plastics aggregation, flotsam effect and fast vertical sedimentation in the oceans (fig.1., side view).

While engaged in cutting and chopping plastics for the marine soup, the conversation led on to how difficult it is for individuals in the UK to shop for daily food that is not wrapped in plastic. From the sea of choices given overall, there is almost no choice for plastic-free products in a UK shopping isle. The workshop conversations ended on an anxious note, as we reflected upon the plastic pollution facts and the decline of the marine habitat. To borrow the words from a participating stakeholder, who regularly follows environmental news, this topic makes her so anxious that "it feels like the end of the world".

Lanterfish—a plastic reuse muse

> *Entanglement is a condition of being twisted together or entwined, involved with; it speaks of intimacy gained, even if it was resisted, or ignored or uninvited. It is a term that gestures towards a relationship or set of social relationships that is complicated, ensnaring, in a tangle but which also implies a human foldedness.* (Nuttall, 2009: 13)

The 'Plastic Soup' workshop also opened up a space to acknowledge the potential of a latent emotional sensitivity to the non-human species living in the ocean. In the oceans, the Plastisphere is creating an instant habitat for microbial life, such as phytoplankton, a microscopic marine alga. The lanternfish follows the migrations of plankton, which serves as its primary food source. We can seek to shift perspectives from the human to that of other species. In becoming empathetically entangled with marine life, I visualise the gyre by following the lanternfish's daily vertical migration, and how it is ingesting large quantities of plastic. This can enable us to access aspects of oceanic life that our postmodernist society doesn't normally see or emotionally relate to: and can change one's worldview. Socio-materially marginalised, nature is politically misrepresented. With no direct rights or voices to be recognised, it falls within the same prioritised, unseen, unheard and hidden category as plastic waste.

Some oceanic species use plastics as pasturage. Marine science reports many interactions between plastic debris and marine organisms. Microbial biodegradation and grazing by macrofauna facilitate fragmentation of plastics (Eriksen et al., 2016: 3.1.1). Macrofauna is any animal visible to the naked eye, such as the lantern fish. The plastic debris entanglement with marine life is well documented, such as the bite marks and ingestion of plastics (Eriksen et al., 2016: 3.1.1). However, the ecological impact and significance of marine life grazing on, or tearing small pieces off plastics, is to date not well understood.

The habitat of the lanternfish species is the mesopelagic zone (intermediate depth), but its night feeding grounds are towards the sea's surface. This small fish (most under 150 millimetres long) dominates the biomass present in our oceans and provides an important food source for many large predators such as dolphins, seals and tuna. Wieczorek et al. (2018: 1–9) report that 73% of all tested lantern fishes contain plastics in their guts. The study found that the highest ingestion rates of microplastics occur in the gastrointestinal tracts of mesopelagic fish. This has important consequences for the health of marine ecosystems in general. If the lanternfish is ingesting plastic, so are we. Thus, the lantern fish is my plastic reuse muse.

Living landfill

"Plastic is wholly swallowed up in the fact of being used: ultimately, objects will be invented for the sole pleasure of using them. The hierarchy of substances is abolished: a single one replaces them all: the whole world can be plasticized, and even life itself…" (Barthes, 1957: 99).

As a petroleum-based society, we have created a space for the birth of a new hybrid environment. The plastic materiality introduced so far by expanding global production will end up discarded, carrying a toxic message into future epochs. As I look upon the oceans, into the Plastisphere a new paradigm of materiality opens. Trying to visualise the gyre from the perspective of a lanternfish helps me, as a human (designer), to make a leap, emotionally and cognitively from my usual position on land, into a much more fragile and unfamiliar space.

Human geographers Steinberg and Peters (2015) propose that their '*wet ontology*' approach can help in thinking '*with*' the sea and assist in reconceptualising the oceans' uncertainty. Through the course of workshops, actions and reflections, participants learned about the North Atlantic Gyre and the forming subpolar Barents Sea Patch and that the lanternfish's grazing habitat is the Plastisphere. In empathising with the other species, I cannot escape shopping single use plastic packaging in the UK supermarkets, and it seems that the lanternfish can't avoid grazing upon the plastics particles either, thus both species are involved with the plastics. Human food is wrapped in the plastic packages and microscopic marine colonies live on the plastic particles. Is the wet and living landfill teaching us, humans, a lesson in circularity and reuse, revealing how human discarded plastic materiality became a living habitat, offering multispecies engagement?

As our planetary environment undergoes radical changes, considering the living conditions of nonhumans, like the lanternfish, needs urgent attention. I feel responsible to participate and support this shift, in terms of worldviews, relations and ways of knowing, in design professions. This design workshop opened to new plastic material relations, and favoured reconnecting with nonhuman species and nature. Thus, the workshop advocated a call to raise a new plastic materiality paradigm and change plastic governance: in why and for whom we relate to, design, manufacture, distribute, consume, dispose and discard plastics.

REFERENCES

— Barthes, R. (1957). *Mythologies*, trans. A. Lavers. New York: Farrar, Straus & Giroux; The Noonday Press.

— Boonzaier, L. & Pauly, D. (2015). Marine Protection Targets: An Updated Assessment of Global Progress. *Oryx*, 50(1), 27–35. DOI: 10.1017/S0030605315000848

— Eriksen, M., Lebreton, L.C.M., Carson, H.S., Thiel, M., Moore, C.J., Borerro, J.C., et al. (2014). Plastic Pollution in the World's Oceans: More than 5 Trillion Plastic Pieces Weighing over 250,000 Tons Afloat at Sea. *PLoS ONE*, 9(12), e111913. DOI: https://doi.org/10.1371/journal.pone.0111913

— Eriksen, M., Thiel, M., Lebreton, L.C.M. & Takada, H.K. (2016). *Nature of Plastic Marine Pollution in the Subtropical Gyres.* DOI:10.1007/698_2016_123. Available at: https://www.researchgate.net/publication/313594097_Nature_of_Plastic_Marine_Pollution_in_the_Subtropical_Gyres

— Hoornweg, D. & Bhada-Tata, P. (2012). *What a Waste: A Global Review of Solid Waste Management.* Washington, DC: The World Bank. Available at: https://openknowledge.worldbank. org/handle/10986/17388

— Jambeck, J.R., Geyer, R., Wilcox, C., Siegler, T.R., Perryman, M., Andrady, A., Narayan, R. & Law, K. L. (2015). Plastic Waste Inputs from Land into the Ocean. *Science*, 347(6223), 768–771. DOI: 10.1126/science.1260352

— KraalD, 2015, Designtransposal workshop, PhD By Design, Goldsmiths University of London, UK.

— MacArthur, E., Foundation (2017). *The New Plastics Economy: Rethinking the Future of Plastics & Catalysing Action.* Available at: https://www.ellenmacarthurfoundation.org/assets/downloads/publications/NPEC-Hybrid_English_22-11-17_Digital.pdf

— Nuttall, S. (2009). *Entanglement: Literary and Cultural Reflections on Post-apartheid.* Johannesburg, South Africa: University of the Witwatersrand Press.

— PhD By Design (2015). *Workshops: Visualising Through the Gyre.* Available at: https://issuu.com/phdbydesign/docs/pbd_conference_documentation_2015

— Steinberg, P. & Peters, K. (2015). Wet Ontologies, Fluid Spaces: Giving Depth to Volume Through Oceanic Thinking. *Environment and Planning D: Society and Space*, 33(2), 247–264.

— Van Sebille, E. & Froyland, G. (2012). Origin, Dynamics and Evolution of Ocean Garbage Patches from Observed Surface Drifters. *Environmental Research Letters.* DOI: 10.1088/1748-9326/7/4/044040. Available at: https://www.researchgate.net/publication/234002610

— Wieczorek, M. Alina, Morrison, L., Croot, P.L., Allcock, L.A., MacLoughlin, E., Savard, O., Brownlow, H. & Doyle, T.K. (2018). Frequency of Microplastics in Mesopelagic Fishes from the Northwest Atlantic. *Frontiers in Marine Science*, May.

— Zettler, E., Mincer, T. & Amaral-Zettler, L. (2013). Life in the "Plastisphere": Microbial Communities on Plastic Marine Debris. *Environmental Science & Technology*, 47(13), 7137–7146. Available at: http://pubs.acs.org/doi/abs/10.1021/es401288x

(5 April 2017) Spring seems early. Swallows arrived yesterday. There is blossom on the plum tree but few pollinators, I'm going out with a paint brush…

Else Skjold, Frederik Larsen

Design and Nature as Seen through Fur: Systems of Manipulation and Care

Fur as system of manipulation and system of care

Fur production is an intriguing, if controversial, space of opportunity with regards to considering design and nature. Fur stands as an iconic example in debates concerning humans and their (negative) impact on nature. In recent times, fur exhibits a relationship between design and nature, which could be seen as a *system of manipulation*. By unpacking this ethical minefield of interrelations between design and nature, we will sketch out how particular animal types and design approaches have been closely intertwined with moral and ethical discussions within Western society up till today. Furthermore, we will show how design aesthetics are used as tool to indeed transform nature into a fabric of industrialism, as traces of individuality are erased through the craftsmanship of furriers.

Fur between industrialisation and nature

As the rise of Haute Couture in the mid-1800s demonstrated the bourgeoisie as the new, ruling class, the material of fur became central as display. The rise of Haute Couture signifies an emerging gender dichotomy, entrenched in the design aesthetics of fur in this period and onwards. Thus, preferred animal types and design aesthetics changed. As described by Skjold and Csaba (Skjold & Csaba, 2018), even if fur is typically considered a luxury material, a look into the interplay between animal types and design aesthetics throughout Western history since the Middle Ages shows how only smaller, short haired animals were considered luxurious, whereas larger animals with longer, coarse hairs were not. In terms of design aesthetics, the smaller animals were typically used as lining on the inside of garments, whereas the larger were used reversely, with the hairs on the outside of the garment. This dividing of animal types and design approaches through the sumptuary laws ensured that the class system was symbolically re-enacted through design and use of fur, as only the aristocratic circles were allowed to wear fur from smaller animals. But from the 19th century and onwards, larger animals with long hair were also made use of in design for the upper classes, such as fox and later mink, or other soft-looking fur types for women, and more 'wild'-looking fur types, such as raccoon, bear or wolf skin, for men (Munnichi, 1992). The hairs were now typically placed on the outside of the garment with the aim of bringing volume to the silhouette, with the inspiration from Russian and East-European folklore (Vassiliev, 2000). Notably, fur more or less disappeared from Western menswear garments from approximately the 1930s, so that from then on and up till today fur became closely connected with Western ideas of femininity. Because of this development, the fur-clad upper-class woman of the 20th century was seen as the spitting image of what de Beauvoir termed 'man's other'; as nature and as body (de Beauvoir, 1989[1952]). In this way fur illustrates and enacts the twinned 'projects' of dominion of man over nature and man over woman.

The use of caged animals which are farmed solely to clad the women of rich men is often viewed to epitomise the way in which we as human race have perceived nature as having only one purpose; to serve us (Olson & Goodnight, 1994). That design of fur has primarily targeted the vanity and aspirations of what Veblen termed 'the leisure class' (Carter, 2003) has contributed to how fur was—and is indeed still—seen as problematic. The debate about fur therefore displays how in the Protestant socio-cultural structure of Western society, luxury consumption has been morally problematised for centuries (Hunt, 1995).

We argue that the relationship between design and nature that fur therefore displays has placed fur as iconic of *systems of manipulation* in recent moral and ethical discussions. The 'Fur Controversy' emerging in the late 1970s Britain addressed fur as intertwined in a line of debates about societal power-structures, and resistance towards fur is still dominating within several oppositional groupings that grew out of the anti-establishment milieu of the 1970s–80s, such as feminism, anti-capitalism and environmental- or other NGO movements, all working towards a world less dominated by growth, power abuse and exploitation of fellow humans, animals and nature (Emberly, 1998).

In the following we will focus historically on the symbolic discourses of fur design aesthetics in Western society, and how they equally align with the industrialist *system of manipulation*. We will elaborate particularly on the in-built dichotomy between nature and design for industrial production of fur in the 20th and 21st centuries. This dichotomy is expressed in the dyeing, cutting, manipulating and stitching together of natural skins into a uniform surface.

Fur as industrialised aesthetic

> *There are two ways in which the grid functions to declare the modernity of modern art. One is spatial; the other is temporal. In the spatial sense, the grid states the autonomy of the realm of art. Flattened, geometricized, ordered, it is antinatural, antimimetic, antireal. It is what art looks like when it turns its back on nature.* (Rosalind Krauss, 1979: 50)

Throughout the 20th century in particular, the design of fur can be perceived as emblematic of a particular industrial aesthetic, as seen through the way the material is being manipulated by furriers and designers. In fur production, the intense process of transforming nature into a workable material for industrialised manufacture becomes visible not only in the breeding for uniformity to meet the demand of designers but also in the intersorting and valuation of skins so that they require minimal processing in the design phase. In order to meet industrial uniformity standards (a key instrument in the modern growth economy), each single skin that arrives at the auction house is subdued to a highly complex intersorting technology that scans for colour variation, animal size and hair length, after which the sorting is quality tested manually.

Alongside sorting to meet the demand of uniformity, there is breeding to meet the demand of choice. For example, whereas the mink animals that were imported from the USA to Denmark in the 1930s (Skov, 2005) were dark brown, today, mink farmers have bred varieties of brown mink, as well as new hybrids such as grey, spotted, black, white, or pearl coloured mink, that all have their distinct name and colour variation. All of these efforts are made in order to ensure that a large variety of skins are available for designers to work with, and that each bunch of skins sold at the auction house represent a seemingly uniform sample of skins. From nature's hand there are 11 various species of mink living in nature—at the Kopenhagen Fur auction house there are presently more than 23,000 categories of mink fur which are intersorted through technology and by hand.

In fur design and manufacture, natural features of fur which are not compliant with certain aesthetics and production requirements are removed or disguised through much intervention. Fur, in order to present as a textile for design, needs to be manipulated and assembled. The finished product, a fur coat for example, is valued for the way in which the furriers are able to disguise the individuality of the single skins, the natural variations. Quite literally, snippets of individual skins are stitched or glued together to form a two-dimensional square unit of maximum 50 × 50 cm to showcase various fur techniques or dyeing methods — in furrier terms named *fur fabric* (observations of furriers in the period of 2014–2018). In the design of full garments based on these samples of fur fabric, there are a number of ways in which the natural features of mink skins are hidden in the design and manufacture process. It involves knowledge of the natural structure of the skins, such as hair length, hair direction, placement of pattern on the animal skin to avoid an uneven appearance of the final design, as well as the skills to transform these features into abstract design elements. One way of disguising natural colour patterns is to create horizontal lines in the design of a coat, in which the design features of a padded jacket are imitated (see fig.). These techniques ensure that the finished design does not bear the trace of the diversity in colour, hair structure and size of individual animal skins. Nature is presented in an idealised form.

Fur as possible learning space for design and nature

What makes fur production and fur manufacture so unique among fabrics is the fact that (leather excluded) fur is not produced by the mechanical processing of fibres into a structured textile, but through a high level of knowledge and skills possessed by craftspeople: furriers. We are interested in how the craft expertise of furriers and the complex relationships between customers, their fur garments and the furriers themselves, can be a space to critique and invigorate a design and nature discourse. Can the connections between a live animal, a craftsperson, a garment and a consumer create a new dialogue?

On the one hand, the creation of fur can be regarded as a system of manipulation in the name of man's dominion over nature and a growth economy. However, the case of fur also offers up opportunities for imagining a different relationship between design and nature, in which both designers and users place a high value on, understand and honour what has been taken from nature.

Certainly, the economic value of fur has promoted the high level of 'care' in fur design, manufacture and use. This is not explicitly related to fur's (and leather's) unique proximity to the sacrifice of a live animal — nature — for design by and for humans. In order to genuinely make a shift from systems of manipulation to systems of care, this unique proximity needs much further exploration and reflection—deeper than an 'easy' emotional rejection of fur on the grounds of class associated morals. What are the opportunities to go deeper into valuing fur and to embrace the full complexity of animal life and death, fur making and wearing? How can the embodied life in fur be cared for? Can this show us the way to care for the embodied life in all of the natural world?

This fur was handed in to furrier Alex Pedersen in the city of Nærum, Denmark. According to the furrier it probably derives from the early 1990s, and is now in the process of being re-designed. In order to mimic the look of a duvet jacket the skins have been placed horizontally, and all traces of the individual animals have been removed.

REFERENCES

- Carter, M. (2003). Thorstein Veblen's Leisure Class. In *Fashion Classics: From Carlyle to Barthes* (pp. 41–59). Oxford: Berg.
- de Beauvoir, S. (1989[1952]). *The Second Sex*. New York: Vintage Books/ Random House.
- Emberly, J.V. (1998). *Venus and Furs. The Cultural Politics of Fur*. London: I.B.Tauris.
- Hunt, A. (1995). The Governance of Consumption: Sumptuary Laws and Shifting Forms of Regulation. *Economy and Society*, *25*, 410–427.
- Krauss, R.E. (1979). Grids, *October*, 9 (Summer, 1979), 50-64.
- Krauss, R.E. (1985). *The Originality of the Avant-Garde and Other Modernist Myths*. Cambridge, MA: MIT Press.
- Munnichi, A. (1992). *Ladies in Furs 1900–1940. The Twentieth Century History of Fashion*. Modena, Italy: Zanfi Editori.
- Olson, K.M. and Goodnight, G.T. (1994). Entanglements of Consumption, Cruelty, Privacy, and Fashion: The Social Controversy over Fur, *Quarterly Journal of Speech*, 80(3):249-276.
- Skjold, E. & Csaba, F. (2018). Fur and Sustainability – Oxymoron or Key to 'Deeper' Luxury? *Journal of Design, Business & Society*, 4(1), 131-149.
- Skov, L. (2005). The Return of the Fur Coat: A Commodity Chain Perspective. *Current Sociology*, 53(1), 9–32.
- Vassiliev, A. (2000). *Beauty in Exile: The Artists, Models, and Nobility who Fled the Russian Revolution and Influenced the World of Fashion*. New York: Abrams.

Erik Sandelin

Bully Goes Fishing

When I was 12 years old I worked weekends all year as a paperboy to save up for it: a tiny orange fibreglass boat with a 4-horsepower outboard engine. It was late March, the best time of year for catching large perch, bulging with eggs, in the river near my home in southern Sweden. Slowly motoring by a group of cows that were grazing next to the river, I scooped up water with the bailer and casually threw some at them. They twitched and mooed and I chuckled to myself.

Planet Earth is no longer a smorgasbord of infinite and readily-available resources for humans to indulge in. Design and technologies called into being by humanist and modernist visions of expansion and economic growth have had unintentional effects that we must now grapple and live with. In academia there are diverse and urgent calls for a dismantling of the anthropocentrism that underpins these visions and actions. Critical posthumanists like Rosi Braidotti (2013), sociologist Bruno Latour (1993), feminist technoscience scholar Donna Haraway (2008) and biologist Scott Gilbert (Gilbert et al., 2012) urge us to rethink ourselves as interdependent beings-in-the-world rather than as sovereign, rational individuals standing outside of the world with full instrumental access to that which is not-us, not-human.

Scholar-activists from Critical Animal Studies (Nocella, Sorenson, Socha & Matsuoka, 2014) and eco-marxists (Soper, 2012; Weisberg, 2014; Malm, 2018) likewise seek to dethrone the imperial human, but they prioritise social struggle over hybrid ontologies and distributed agency. We (capitalist and speciesist) humans are the problem. We are the ones who are responsible and capable to organise a resistance against fossil capital and the animal-industrial complex.

For designers, these more-than-human predicaments present substantial challenges and opportunities. How can we adapt design processes to involve and account for more-than-human stakeholders and concerns? How can we put designerly skills to use in making manifest new, less violent relations with fellow critters?

In a recent lecture titled *The Regenerates* Chrisna du Plessis (2017) urged her audience, primarily PhD students in architecture and design, to go "beyond sustainability" towards "regenerative design". As sustainability has come to mean minimising harm while maintaining smooth economic operations, rather than transforming the world for the better of all creatures, du Plessis encouraged positive thinking and to "reconnect with nature".

I was annoyed, and I couldn't fully understand why. I agreed with the need for reconfiguring human subjectivities and desires for lasting change to take root. I had some gripes with her use of the concept of "nature", but what I felt was a more intimate, even childish, aversion. When du Plessis acknowledged that her call to reconnect with nature and "the heart" might seem "a bit twee" I understood my grudge. "Twee", a neologism supposedly born of a child's mispronunciation of the word "sweet", is an excessively quaint style of pop music. If The Regenerates was a band I would not be a fan.

In *The Ecological Thought* Timothy Morton writes, "[m]aybe at heart I'm an old-fashioned goth, but when I hear the word 'bright' I reach for my sunglasses". Morton stresses the profound uncute-ness and uncanny-ness of ecological thinking. "Environmental rhetoric is too often strongly affirmative [...] It's sunny, straightforward, ableist, holistic, hearty and 'healthy'. Where does this leave negativity, introversion, femininity, writing, mediation, ambiguity, darkness, irony, fragmentation, and sickness?" (Morton, 2010: 16). Ecological thinking and design must include gloom as well as light, negativity as well as positivity.

In this text, I suggest that design reaching beyond the human should foster a sensibility also towards these darker timbres of human–nature interactions. Starting from the silly-yet-profound feeling of not digging the bright song that often plays I try another tone of voice, through the figure of Bully—a troubling-yet-hopeful character who could have a role to play in design practices that aim to be less anthropocentric. Bully takes shape through three experiential vignettes and reflections on design acts that reconfigure interfaces to non-human lives: genetic manipulation, meat-eating and recreational fishing. Who is Bully and what can he do for us? Let's see.

I'm in a biotechnology lab in Helsinki, wearing a white coat, bulky safety goggles dimming my vision, my hands latex-gloved. In the petri dish on the bench are colonies of Escherichia coli bacteria. The dish is still warm from the incubator and smells like alien cheese. The bacteria are now "competent", their cell walls permeable and about to absorb the foreign DNA at the ready in my pipette.

At the Merry CRISPR workshop in December 2017 a small group of artists and designers explored novel gene editing technologies that could permanently alter the DNA of cells in living organisms. After successfully transforming a strain of *E. coli* to express a gene that made them resistant to a particular antibiotic, my group sat down and talked. We talked about the way we had violently force-fed synthetic DNA to the bacteria. It had been a power trip. We had enacted violence on a microscopic scale but violence and control still characterised our relation to these living beings — as it does to the vast majority of nonhuman animals on the planet. We talked about other possible relations and drew a spectrum on the whiteboard. The two opposite poles were occupied by Angel (the pure abolitionist with clean hands) and Killer (the human supremacist). Who were in the middle? Diplomat, Trickster, Manipulator, Bully, to name a few. Bully asserts his position by intimidating those with lower status. What would Bully do in the lab? We started talking about quorum sensing (Whiteley, Diggle & Greenberg, 2017), the way bacteria communicate and organise themselves using chemical signals. Bacteria can sense how many they are in a population to coordinate joint endeavours, such as enabling bioluminescence in a squid or launching a virulent attack, when there is a critical mass. By understanding the social life of bacteria we can potentially learn to intervene by adding noise to the bacterial conversation. Biotech-Bully would enjoy manipulating and confusing the bacteria, but the use of lethal antibiotics or intrusive genetic operations would be out of bounds. Bully is not Killer.

Garfish caught (on camera). Photo: Erik Sandelin

I'm standing on a rock at the end of a pier in southwestern Malmö. A few other fishermen are repeatedly casting and winding in their silvery lures through the brackish waters of the Öresund strait. It's a warm May morning and we all know that schools of slender garfish have migrated from the Atlantic closer to the coast to spawn. Suddenly I feel a distinct tug on the line. Instinctively—a reflex from 30 years ago—I pull up on the rod with a quick jerk. As I didn't intend to fight a fish today I'm relieved when I feel the line immediately slacken.

As a teenager I would spend countless days by the river patiently waiting for a large pike or perch to bite. When I stopped eating animals 15 years ago I also lost the passion for catching fish with a hook. I do miss being outdoors, fiddling with the fishing gear, the convivial chatting, the suspense in waiting for a bite, the tiny triumph of a catch. In a recent experiment, inspired by the Bully character of my illustrated storybook *Designer and Goldcrest* (Sandelin, 2018), I tried an alternative setup to retain these qualities of traditional recreational fishing without murdering or piercing a single fish. Using a hookless lure and a small underwater camera attached to the line I can go "fish watching". I get to trick the fish and "catch" them on film. I'm ok to act as Bully, but not as Murderer.

The underwater camera used is promoted (Kellerman, 2014) with an unusually frank celebration of the pleasures of human supremacy in fishing:

> *There are many things to enjoy about fishing. There's the communion with nature. The taste of something pulled right out of the water. The satisfaction of providing food for yourself from the wild. But obviously, what's missing is the satisfaction of watching stupid, stupid fish ensuring their own demise. That's right fish, who's got a fully developed cerebral cortex now?*

For Bully this would be going too far. Bully may be a voyeur, but not a sadistic killer. But the technology can help bridge activities with different ethics. So far I'm socially accepted by the other fishermen on the pier. "How much does the camera weigh?", one man asked me. We can talk about the technology, as my rig has the same appearance as his. I thought of how the vegan at the shared Christmas table can use tofu sausages and mock-meatballs to defuse potential social friction (Nath & Prideaux, 2011).

We were carefully attaching the creature's oven-baked gluten arms but something felt wrong, disrespectful. It wasn't that E.T. was skewered on a steel rod, from hip to mouth, ready for the rotisserie. I knew "he" wasn't real—just a lump of bread, really. But when we tied the arms in place with steel wire behind his back, exposing his wrinkly belly, I knew this wasn't the way to treat an extraterrestrial visitor, even one built from wheat. We had to tie up his hands properly before the barbecue, not as a prisoner to be humiliated, but as a dignified source of protein.

In May 2014 a full-scale gluten replica of E.T., a fictional alien from Steven Spielberg's 1982 Hollywood blockbuster, was roasted whole on a spit and eaten at a public barbecue in Lund, Sweden (Sandelin, 2014). The project was a collaboration between Unsworn Industries and artist Terje Östling. Several of the participants at the barbecue were hesitant to carve into, let alone eat, a creature (albeit fictional) with which they shared a long-standing emotional attachment. Although contemporary carnivores may only rarely experience a sense of loss or attachment when grabbing a hot dog from the local kiosk, there is undoubtedly a sacrificial aspect to meat eating. Cary Wolfe (2013: 97) asks: "Does sacrifice make meat taste better? Indeed, is it what makes meat 'meat'?" The gluten E.T. did taste a bit leathery, but when we enact the sacrifice of consuming the plant-based extraterrestrial, does that make us feel more "human" than gobbling down a seitan sausage or even an in vitro meat burger? Perhaps the Mock Alien caters to those underbelly-qualities of meat eating: domination, power, sacrifice—qualities that would earn barbecued E.T. a place on Bully's menu.

* * *

As we move from *ego* towards *eco*, foreground and background start to overlap and blur, writes Morton (2010). Without the comforting contrast between Us and Them, between subject and object, we inch closer to a schizophrenia, everything potentially meaningful and meaningless at the same time. On the contrary, Malm (2018: 77) argues that the contrast between nature and society, them and us, will become starkly tangible as "nature comes roaring back into society in a warming world".

Whether dizzying interdependence or a quickly approaching, roaring storm—these are painful transitions. Rosi Braidotti (2013) faces the pain with joy and affirmation. Malm (2018: 16) embraces anger, urging for direct action. It's time to "deflate the tyres, block the runways, lay siege to the platforms, invade the mines".

Who is Bully? Bully is neither a paragon of ecological interconnectedness nor a militant activist. In the biotech lab Bully empathises with the bacteria and seeks creative alternatives to genetic penetration. Gluten E.T. is a source of protein and artificial mastery for vegan Bullys. Well-crafted surrogates may suffice to satisfy Bully's needs. Bully finds pleasure in asserting his position by kicking downwards, but not too hard. Fish-watching-Bully is a user of a redesigned practice that maintains the excitement of the "catch", with less bloody trophies. Bully is a persona with a tainted ecological register, designer or user.

With bullies currently occupying some of the world's most powerful positions my intention here is not to celebrate bullyish behaviours. It is to remind designers that our practices require a keen ear for both bright twee pop and blackened doom metal; a palette of aesthetics and timbres that acknowledges the many humanoids who tread the murky waters between angelic, abolitionist absolutism and business-as-usual human supremacism.

REFERENCES

- Braidotti, R. (2013). *The Posthuman.* Cambridge, UK: Polity Press.
- Gilbert, S.F., Sapp, J. & Tauber, A.I. (2012). A Symbiotic View of Life: We Have Never Been Individuals. *The Quarterly Review of Biology*, 87(4), 325–341.
- Haraway, D.J. (2008). *When Species Meet.* Minneapolis, MN: University of Minnesota Press.
- Kellerman, J. (2014). *Watch Fish Fall for It with Strike-Cam.* Wide Open Spaces. http://www.wideopenspaces.com/watch-fish-fall-strike-cam/ [accessed 21 November 2017].
- Latour, B. (1993). *We Have Never Been Modern.* Cambridge, MA: Harvard University Press.
- Malm, A. (2018). *The Progress of This Storm: On Society and Nature in a Warming World.* London: Verso.
- Morton, T. (2010). *The Ecological Thought.* Cambridge, MA: Harvard University Press.
- Nath, J. & Prideaux, D. (2011). The Civilised Burger: Meat Alternatives as a Conversion Aid and Social Instrument for Australian Vegetarians and Vegans. *Australian Humanities Review*, issue 51.
- Nocella, A.J., II, Sorenson, J., Socha, K. & Matsuoka, A. (2014). *Defining Critical Animal Studies: An Intersectional Social Justice Approach for Liberation.* New York: Peter Lang.
- du Plessis, C. (2017). The Regenerates. Lecture delivered to PhD Course ResArc Approaches at Chalmers, Gothenburg, 7 December 2017.
- Sandelin, E. (2014). Eating E.T. In E.A. Cederholm, A. Björck, K. Jennbert & A.-S. Lönngren (Eds.), *Exploring the Animal Turn: Human–animal Relations in Science, Society and Culture* (pp. 47–56). Lund: The Pufendorf Institute for Advanced Studies.
- Sandelin, E. (2018). *Designer and Goldcrest.* Malmö: Sorgenfri Press.
- Soper, K. (2012). The Humanism in Posthumanism. *Comparative Critical Studies*, 9(3), 365–378.
- Weisberg, Z. (2014). The Trouble with Posthumanism: Bacteria are People Too. In J. Sorenson (Ed.), *Critical Animal Studies. Thinking the Unthinkable* (pp. 93–116). Toronto: Canadian Scholars' Press.
- Whiteley, M., Diggle, S.P. & Greenberg, E.P. (2017). Progress In and Promise Of Bacterial Quorum Sensing Research. *Nature, 551*(7680), 313–320.
- Wolfe, C. (2013). *Before the Law: Humans and Other Animals in a Biopolitical Frame.* Chicago, IL: University of Chicago Press.

Mathilda Tham

Dirty Design (or A Bloody Mess) *In Celebration of Life Affirming Design*

Making notes for Design and Nature (2018), watercolour on paper by Mathilda Tham

It's a peaceful August morning, smells of dry earth—2018 an unusually (ominously) hot and dry summer, sounds of my pottering daughters, the gentle breeze through the trees, the rustling of birds and the occasional squirrel. It's the sixth morning of camping in the woods, my bones habituated to sleeping on the floor and my mind to the dip in hygiene. This morning the world is soft and clear and easy to be in, my skin also easy to be in. Actually, today the membrane between me and the world feels unusually thin. I do feel *part of*.

Yet, as I'm lying in the hammock, sketching and making notes for my piece on Design and Nature, I realise how much the human gestures and designs help me to appreciate nature, this forest. The hammock, shelter, fold up chairs, fireplace, teapot and cafetière—the clearing itself—anchor and orient me. These human designs frame—tame—my encounter with nature, making it acceptable, safe and orderly to me. Idyllic at first, the scene—surely, a simple human–nature encounter—is further troubled by the lab coat I'm wearing (I do when I paint, even in the woods) as well as one of my daughters' lab coats draped over a chair.[1] The lab coats epitomise how I'm not prepared to meet nature naked. I meet her dressed in scientific armour, emphasising not only the distance that the gear/the design creates between me and her, but also the position of power I take in this distancing. Here, I follow in the steps of scientific revolutioneers such as Francis Bacon, one of the first of a line-up of men dedicated to the sentiment "By art and the force of man… [nature can be] forced out of her natural state and squeezed and molded… [and] human knowledge and human power meet as one" (Bacon as cited in Merchant, 1982: 17). In the woods, in this armour, it becomes so clear to me how design is not only a part of a scientific heritage of colonising and taming nature, even killing life (and of course also saving human lives), but profoundly a precursor. Design is not just 'One of many fields' And Nature which Routledge might offer a book on. Design is the instrument that distinguishes humans from nature.

I also reflect, noticing the ant approaching my toe, that this is linked to a fear of life—biophobia; the designed tools allowing the use of nature as resource and creating a safe distance from nature. This makes me feel sad and ashamed —me too Brutus. I would like to be a designer and human that celebrates life. In this text, I wish to explore fear of life and some pathways to life affirming design.

Fear of life

It's true that I'm scared of life. And I do know what happens if I don't get in there first with my design instruments and tame nature. Brambles grow wild. Hair grows wild. Breastmilk stains my clothes, and blood does too. Mould tickles slow food habits. Tiny organisms dance across our table, with the mice and the head lice (after the recent headbanging at dinner time). They dance straight into our bodies, as confirmed by yet another message from the school; this time it's ringworm. It's a bloody mess and I could murder for some peace and quiet, as this abundance of domestic biodiversity and involuntary signs of life coming from my own body is killing me. It scurries, it trickles, it scratches and tickles. I feel out of control.[2]

I've noticed that management people are scared of life too. However, now we're scaling up from domestic biodiversity and the fauna, flora and fungi found in places of recreation, to the very nature of nature: its unpredictability, messiness, unruliness — complexity.[3] At yet another conference, yet another suited man, mansplains to his audience that the world is complex. Not so much before, but now due to (manmade) technology and (manmade) environmental and social damage it's really bad. But fear not, the talk ends with a pitch for a new tool that this man, our hero, has designed. It is The Superior Management Tool that will be The Final Solution to Managing Complexity (at least until we hit complexity 0.3) and it fits on just three PowerPoint slides (even considering the somber, yet potent executive logo at the bottom of the slides).

Now probably is the time to say that I recognise that management people come in all shapes, sizes and genders; that there is one lurking inside me, as probably in many of you, at work and at home. Here, I want to discuss the thread of noninnocence running from the designed artefact to designed models. The tangible design tool was and is used to avail humans of natural resources at an increased rate and scale (which other species cannot keep up with) and protect humans against other scary species. This has created an inevitable power imbalance, and both a planned and resulting distance between humans and other species. Through this separation, species other than humans become 'nature' and 'other'. The intangible tools of the, so called, knowledge economy do the same thing at increased scale and rate. Here though, the motivating fear concerns not the ferocious teeth of the wolf, but the abyss of not understanding and not being in control.

The 'noninnocence' of care that Maria Puig de la Bellacasa (2017: 164) draws our attention to — the activity of care's inseparability from, for example, power hierarchies — is true also for design. As I sit drawing on a fine August day, my pen traces certain ways of seeing, ways of knowing, ways of doing — and not certain others. Harshly, design, in its immediate and expanded senses, is a war against ontological and epistemological diversity, just like on biodiversity itself. The sharp point of my pen, the sharp blades of Gillette, the sharp outlines of a matrix, are distant relatives of fascism, there to cut out diversity, entanglements, mess, the colourful, the loud in the 'wrong way'. Fear of diversity is embedded in the modernist project and in modern design—'design as hygiene'. Order and method, tools and strategies need to be researched, developed, implemented, evaluated and benchmarked so that finally we can gain control of that unruly life—whether it's the errant sleeping habit of the young child, the social dynamics of a workplace, the messy kitchen cupboard—and get a life. Oh, the solace I take from a good clearing. Oh, the efforts we have made to keep the life, complexity, emotions, confusion at bay.

Dirty design doing—life affirming design practices

So design is a scalable tool that can kill not just nature, but the nature of nature. But of course, just like each autumn term sees a return of the headlouse, 'not understanding', 'confusion', 'not feeling in control' always come back. Is there another way for design than to be the tamer, the settler, the filler of voids, the pro-actor, the pro-truder and in-truder through instruments, the eradicator of confusion, the route to control? Is there another way to relate to all this life, all this complexity?

On a train ride through southern Sweden my six-year-old daughter is bored, so I give her my headphones and phone and plug her into the opera Carmina Burana. Ten minutes into the music, Paloma takes off the headphones and says elatedly to me: Mamma, it's chaos! Then she puts the headphones back on and continues listening to the opera, content to be in the midst.

Excitement and joy are possible reaction to chaos, entanglements, mess, complexity, uncertainty. I guess, just acceptance to be in the midst will also do. Let's just hack it, complexity is here and it's not going away. What's newish is the idea that we can kill it. As many times before, I turn to Donella Meadows' text "Places to Intervene in a System" (1997) and reflect on how, as so often when facing systemic issues, we try to solve them by 'numbers' and by adding 'buffers'. We respond to the students confused by a complex course by giving more and more information, usually leaving them more overwhelmed. Instead we might celebrate with them: "great, you are in search of clarity", or better "great, we are in search of new pathways together". We might remind students and ourselves to leave the big models and frameworks and even tools for now, and instead start digging where we stand and get dirty. It is so clear in design that the genuine love of a project comes when we start digging deep, with our bare hands, getting our psyches and souls in there. Life does not reveal by proxy. Puig de la Bellacasa (2017: 20) emphasises the significance of such physical action for committing to relations, and the need to reclaim touch, a "proximal intimate knowing". "[Touch's] unique quality of reversibility, that is, the fact of being touched by what we touch, puts the question of reciprocity at the heart of thinking and living with care".

Important questions for the designer become: How did you keep yourself dirty today? What did you touch and were touched by? How did you learn amongst the trickle and tickle, the scurrying and scratchy?

Dirty design knowing—life affirming design epistemologies

My four-year-old hands me the mobile phone and says, 'speak to mormor!'. I take the phone and I speak, first gingerly then with growing confidence and tear blended pleasure. My mum has been dead for eight years and two-hundred and five days now so my daughter never met her. Yet, she presences her grandmother according to her needs, and she presences her to me, possibly as she interprets my needs. My mother, as Rosa Lulu conjures her up, brings wonderful lollipops; to me she brings a warm breath through time and the imprint of a loving gaze, otherwise lost to me. Rosa Lulu is entirely agile in weaving ways of knowing together.

Epistemological agility, diversity and equality are ways of keeping design dirty and life affirming. A principle of metadesign is to support synergy in collaborations since, when faced with complexity such as climate change, no one individual or organisation can hold neither problem nor solution (Wood, 2007; Tham, 2014). Complexity must be distributed across many knowledge holders. Similarly, knowing in complexity must also be distributed across many ways of knowing, and across multiple senses.[1] Only epistemological diversity and abundance will enable design to respond from within complexity, life and nature.

The innate dirtiness of design knowing, which comes from design resting on both artistic and scientific grounds, we must cherish. From my experience, it takes guts to stand up for sensual knowing or emergent processes in transdisciplinary collaborations. Next to precise measurements or a theoretical argument, the subtle resonance of a colour or a mood can appear insignificant as well as threaten to break the cleanliness of an unambiguous table.

But we must insist on these weaker signs of being *part of*—for life, for nature. We can remind ourselves to trust many senses, and that it is also ok not to know. "Knowing is not about prediction and control but about 'remaining attentive to the unknown knocking at our door'" (Deleuze, 1989: 193 in Puig de la Bellacasa, 2017: 91).

Genuinely caring for collaboration and for many ways of knowing, also means caring for the 'awkward spaces' (Jones, 2012) that come from unfinished sentences, and daring to show something less than perfect. This is risk-taking with a small r, and its fumbling qualities need to be nurtured by every design school and company. The always steady hand does not care for life. Life affirming design knowing also requires scrutinising our implicit and explicit alliances with sectors, epistemologies, methods. Design can form new dirty contracts, challenging, for example, the purity of a serving a growth economy.

Dirty design thinking—life affirming design ontologies

> *Thinking in the world involves acknowledging our own involvements in perpetuating dominant values rather than retreating to the sheltered position of an enlightened outsider.* (Puig de la Bellacasa, 2017: 10)

In Carolyn Merchant's (1982) twinned history of women and ecology, the fatal blow causing 'the death of nature' is attributed to the Scientific Revolution. Now we're talking paradigms, ontologies and worldviews, and of course the license that the notion of 'death' has given humans to continue exploitation of nature. Recently though, I can feel a tingling in the outer limbs of a nature and design body, promising awakening. A stretch, a yawn, a rustle. Nature is not dead, she's just waking up from a long sleep in our minds. A more vigorous stretch. And we are part of her. The dirtiness of awakening lies in the death of separation; the death of categories and dichotomies. Such scaffolding has directed our gaze away from the uncategorisable, gritty and fractious. Now this is part of our field of vision; it's soundly blurry and dirty. Dirty design is searching. When asked a question, it no longer says 'it's this' and 'it's that'. Instead it says, 'it depends'.

> *'it depends': it's a nuance that accompanies the ways acts of care are realized, and this again is affected by relational constraints—requirements of an ecology, obligations of the practitioners and their struggles.* (Puig de la Bellacasa, 2017: 164)

Dirty design does not say 'ta da', in an instant revealing the shiny solution to the surprised spectators. Dirty design sees no such surprise as there are no spectators, only co-searchers who have been part, and will continue to be part of unfolding. And dirty design sees no heroes.

Life affirming dirty design

"The point is to make a difference in the world, to cast our lot for some ways of life and not others. To do that, one must be in the action, be finite and dirty, not transcendent and clean". (Haraway, 1997: 36 as cited in Puig de la Bellacasa, 2017: 88)

Swimming weightless in the salty water, crying with my dying mother, breathing cheek to cheek with my baby daughter, making love, tasting the sweetness of the first strawberry of the season, feeling the vibrations of my voice inside the multitude of voices of the choir. In those moments, I felt most alive and I was never so afraid of losing my life. In design, we also know the experience as 'flow', an intense feeling of presence (Csikszentmihalyi, 1990). It's so much easier to conjure up commitment to life and dirt in such moments of drama than on a reasonably ok day in front of my computer. The doubt is in the distance, not in the embrace.

It is in the being apart that being part appears hard. Crossing the threshold to life affirming design practices—such as choosing to tune into collaboration instead of conflict or heroics, remembering the full spectrum of colours when the client prefers white, spending time with a tree when the clock is ticking—seems not worthwhile (as hard as keeping up the running on a rainy day). There is also a whisper of fear: "what if life affirming design means I will become an indistinguishable raindrop, unmemorable, without status, even unlovable? Maybe I do need to stand on this other life, on nature, to be me, a designer, somebody."

We need the continuous, humble practices tying us to the soil, the dirt, to remember the abundance of being part. Puig de la Bellacasa gives the example of "composting—a relational practice that engages ways of knowing' (2017: 147). 'Ongoing doings thicken the meanings of the principles" (2017: 146), and even the small experience of agency opens up new doors for learning, relating, doing and sharing (Tham, 2008). They are threshold practices; pivoting us from outsider to insider and from cynic to engaged (Tham, 2008). Again, this is a turn from the design heroism of modernity—in turn inseparable from a Western hegemony and patriarchy—towards a feminist ethics of care, and a politics (and opening up for a biopolitics) of the mundane and everyday (Puig de la Bellacasa, 2017 citing Jaggar, 2001).

There is also solace, different from the solace of a good clearing, to take from good dirt. Accepting our place in "a web of living co-vulnerabilities" of all life (Puig de la Bellacasa, 2017: 145), turns the valuing of me and my design away from the gaze of humans, and with a soft breath of relief, I am free to imagine where. It's still a beautiful August morning when I close my notebook and close my eyes, letting other senses guide me. I take courage from the wise words of my partner Duncan, as he after at least a decade of fractious coexistence with my field of work, decided on a stance: "It's just metadesign, innit". And I sing, sensing a resplendent ant, bird, squirrel, soil sing-along chorus, "it's just the universe, innit".

NOTES

1 Admittedly, I am a control freak, keeping my daughters stain free when they paint too AND keeping aesthetical control of the stainfreeness measures.

2 And here, Orwell's (1945) notion that "all animals are equal, but some animals are more equal than others" has never been truer, as I frequently and fearlessly peruse my local health food shop to support the quality of life of the bacteria of my gut.

3 Although, for all I know management people are scared of mice as well.

4 See also an extended epistemology in action oriented research, e.g. Heron and Reason (2001).

REFERENCES

— Csikszentmihalyi, M. (1990). *Flow: The Psychology of Optimal Experience.* New York: Harper & Row.

— Heron, J. & Reason, P. (2001). The Practice of Co-operative Inquiry: Research With Rather Than on People. In P. Reason & H. Bradbury (Eds.), *Handbook of Action Research: Participative Inquiry and Practice.* London: Sage Publications.

— Jones, H. (2012). *Practising Awkward Space in the City.* PhD Thesis, Goldsmiths, University of London.

— Meadows, D.H. (1997). Places to Intervene in a System. *Whole Earth Review,* issue 91.

— Merchant, C. (1982). *The Death of Nature: Women, Ecology and the Scientific Revolution.* London: Wildwood House.

— Orwell, G. (1945). *Animal Farm.* London: Secker & Warburg.

— Puig de la Bellacasa, M. (2017). *Matters of Care. Speculative Ethics in More Than Human Worlds.* Minneapolis, MN: University of Minnesota Press.

— Tham, M. (2008). *Lucky People Forecast: A Systemic Futures Perspective on Fashion and Sustainability.* PhD thesis, Goldsmiths, University of London. http://research.gold.ac.uk/11301/

— Tham, M. (2014). Creative Resilience Thinking in Textiles and Fashion. In Jefferies, J., Clark, H. and Wood Conroy, D. (eds.) *The Handbook of Textile Culture.* London/New York/Sydney: Bloomsbury.

— Wood, J. (2007). *Design for Micro-Utopias: Making the Unthinkable Possible, Design for Social Responsibility.* London: Ashgate.

SECTION IV: WALKING

Walking to move

As we walk, with each step we simultaneously rise up and anchor down. Propulsion, it turns out, requires grounding. Walking is a process both of co-ordination and connectivity. It involves alternating between touching earth and rising up and putting that process in motion in life. This final section of the book brings together eight contributions that have an active, dynamic quality about them, sometimes involving locomotion, and brings to the forth a practice of design that is physically active and typically located in a range of real world 'moving' contexts. The significance of walking as a pedagogy, a research process, a philosophical and spiritual practice and a creative process for design, is well documented in scholarly literature. Walking, using mind and feet in concert, is perhaps the oldest method for developing new knowledge, including knowledge about design. The following contributions explore design and nature through a fusion of action and experience.

Walking opens with an essay that is at once both rich and pragmatic by Zach Camozzi. Zach introduces a method to deepen connections with nature, 'Earthbond Prototyping'. In it, he describes a design process that takes its practitioners outside both in a literal sense, that is beyond the controlled environment of a studio, and more figuratively, outside the familiar conventions and formulations of design work.

Timo Rissanen and Laura Sansone also focus on educational methods in their piece exploring experiential learning and teaching approaches employed in a first year undergraduate course about climate change in Parsons School of Design in New York City. Their methods, in a hyper-urban setting, engage with a sense of nature's awe, responsibility, care and resourcefulness through inspired use of potted plants, flower market rejects and food waste.

Also from North America, although this time further west and north, Cameron Cartiere and Nancy Holmes, in their essay on collaborative work with nature, examine their experience of working with bees. Theirs is a tale of constantly adapting processes and culminates in the finessing of a practice, evocatively described as *letting nature lead.*

For Clara Vankerschaver, following nature's lead has seen her enlarge her textile design practice to encompass farming as she grows the experimental fibre crops she uses in her work. Cultivation, for Clara, is defined as "a more developed and improved awareness", an understanding grown in her experience of natural cycles and in the physical manipulation of the fibres themselves.

Maarit Mäkelä shares insights from a period spent in New Zealand, where she would walk to familiarise herself, explore and find materials for her craft, ceramics, such as the black sand from a volcanic beach. The ground, and the character of the specific place, becomes a key agent in the emerging artefacts, creatively challenging the relationship between human and designer and nature.

Providing a shift in epistemological scale, Alyssa Schwann explores Traditional Ecological Knowledge and Wisdom of Canadian Indigenous communities about land stewardship in order to renew a shared relationship with nature. Her insightful piece is predicated on a fundamental readjustment of the way that land is viewed and suggests that design must respond to the wisdom of indigenous elders who know the place, to the roots of a location, that which is "sprung from the land".

Sarah Pennington and Eleanor Margolies maintain this connection with place in their finely conceived and written essay, Antarctica SE3. In it the pair describe a process of taking facts about climate change and "thickening them with care". They do this via an expedition on foot between three early Polar explorers houses' in south east London and unfold insights that speak powerfully about the role of nature and design practice of participatory engagement, maintaining multiple perspectives and embodied experience.

Trading London for the Peak District area of the north of England, in the final contribution in *Walking*, Kate Fletcher unfolds stories of integrated experiences of design and nature. Her tales, which embody a mutualistic practice of design are rooted in interrelatedness and bring together a den, the rain, flowers and a broken fashion system in order to intermingle human interests with all others.

… And we've had broccoli and rhubarb for a few weeks now. Salad in the greenhouse. My nose and cheeks have gone a little brown.

Zach Camozzi

Earthbond Prototyping, a Method for Designers to Deepen Connections to Nature

Earthbond Prototyping is an interdisciplinary and experiential method that includes aspects of design (making), outdoor education (inquiry) and self-reflection (autobiographical writing) that enable designers to learn and share connections between design, people, things and the natural world. It supports a change in the ecologic awareness of the designer, leading toward an ethical perspective on nature within design practices.

I grew up in a small coastal town near Vancouver, Canada. Exploring the forest and ocean was where I was most comfortable. I believe this curiosity is why I became a maker and planner and at that stage in my life it led me into engineering. There I was slowly pulled away from the physicality of my youth toward the dense workforce of mechanised production. Through it all I kept falling back into outdoor pursuits, slipping away during the summers to go canoeing and run workshops on building boats, fires and shelters. These forays into outdoor education sparked my curiosity about design and its varied approaches. So I joined Emily Carr University of Art and Design expecting the holistic and moral aspects of outdoor education to continue to influence my path. In some ways Earthbond Prototyping, developed during my masters of design at Emily Carr, allows others to tap into a method of designing I had been pursuing my entire life. One that allows for designs that "cultivate a deeper sense of connection and obligation to nature" (Orr, 2002: 23)

To engage with the methodology designers must move away from the studio or classroom. These familiar settings are convenient for their controlled temperatures, lighting, air movement, smells and even access by others but they are the box that we so often talk about working and thinking outside of. It is difficult to recognise what we might be missing when working in controlled environments and its effect on our practices. Earthbond Prototyping defines a strategy to immerse oneself in design and nature to discover what this might be.

The evolving characteristics of Earthbond Prototyping are:

1 Seek the unfamiliar

Place yourself in an unfamiliar outdoor setting. This does not need to be a remote setting. Many of us have unfamiliar settings within a short walk from our homes. Simply walk off the trail in a local park, spend extended time in places that you often breeze through or go to a location at night that you normally only visit during the day. You will know it is unfamiliar when it feels a bit uncomfortable. Maybe your arm hairs rise, your sweat evaporates in a cold wind or you are in awe of the scenery (fig. 1).

Figure 1: Early Mornings.

2 Engage with nature holistically

Learn to perceive the world you are immersed in and experience your surroundings with all senses (Chambers, 2008: 122). Be patient and start by picking a spot to do nothing physical for a time. Be present and take in content from as many perspectives as you can imagine, open yourself, be still, be silent. Close your eyes and listen to the sounds around you, smell the air, feel the moisture on your skin. Does sunlight heat a small portion of your arm as it filters through the branches? Flit then focus on a sense, how many layers of information do you notice? Do not pretend that as an observer you are not influencing the environment; your heat transfers to your surroundings and your form changes wind patterns.

Once you feel comfortable, move within this place. Touch and consider textures, terrain, balance and relation. Classen reminds us that "while the eyes are trained how to see, the hands should not be kept "blind", grasping after knowledge which is withheld from them" (1999: 274).

You might be tempted to separate your thoughts from senses and feelings but remember that this is impossible. The moisture you hear dripping on a leaf may be the reason why you are grimacing at the clouds looking for signs of rain. Pay attention because the goal, through practice, is to invite understandings of sentient and non-sentient relationships, not to remove oneself from the place by imagining. This means you may have to zoom into fine details to catch ants harvesting the nectar of aphids, lie on a stony shore to notice the sun's heat dissipating from Agate and Jasper or just be present for an extended period of time to observe that your grassy backyard is actually shared with woodpeckers, bees, slugs, crows, ladybugs, squirrels, rats and occasionally your neighbour's cat. Embracing these moments alters perceptions and can change how we act. Heesoon Bai says this well: "our actions in the world emanate from who we are: the way we think, perceive, sense and feel" (Bai, 2012: 312).

3 Loosen control

Take a moment to realise that design, regardless of the context, is often about control: we manipulate materials and environments, discover and define constraints and shape new experiences through artefacts. Putting yourself in an unfamiliar setting strips some of this away. If you are feeling apprehensive, playfully engage with the environment around you. Push through bushes, climb a tree, remember what it is like to be a child and take off your shoes. This playfulness reduces inhibitions and our fear of risk; ruining a shirt, falling from a height or stepping on something sharp is validated by the exploration. You may just find yourself back at step one seeking another unfamiliar setting.

4 Make with self-originating materials and phenomenon

Within the unfamiliar and shifting context, develop a prototype with what nature supplies. Make with natural materials such as sticks, dirt and leaves or phenomena like wind, rain and sunlight. You'll find the wind direction, time of day or topography shifts your objective; ideas will emerge in how the environment interacts. Build with sand to shape the incoming tide, create an image of shadows or use sap as a binder to form build with debris. This experimentation with controls leads to understanding what material properties afford but also gives nature the time to be a design participant. At this stage your creations will not mesh seamlessly with nature's processes but they will expose them over time (Van Der Ryn & Cowan, 1996: 74).

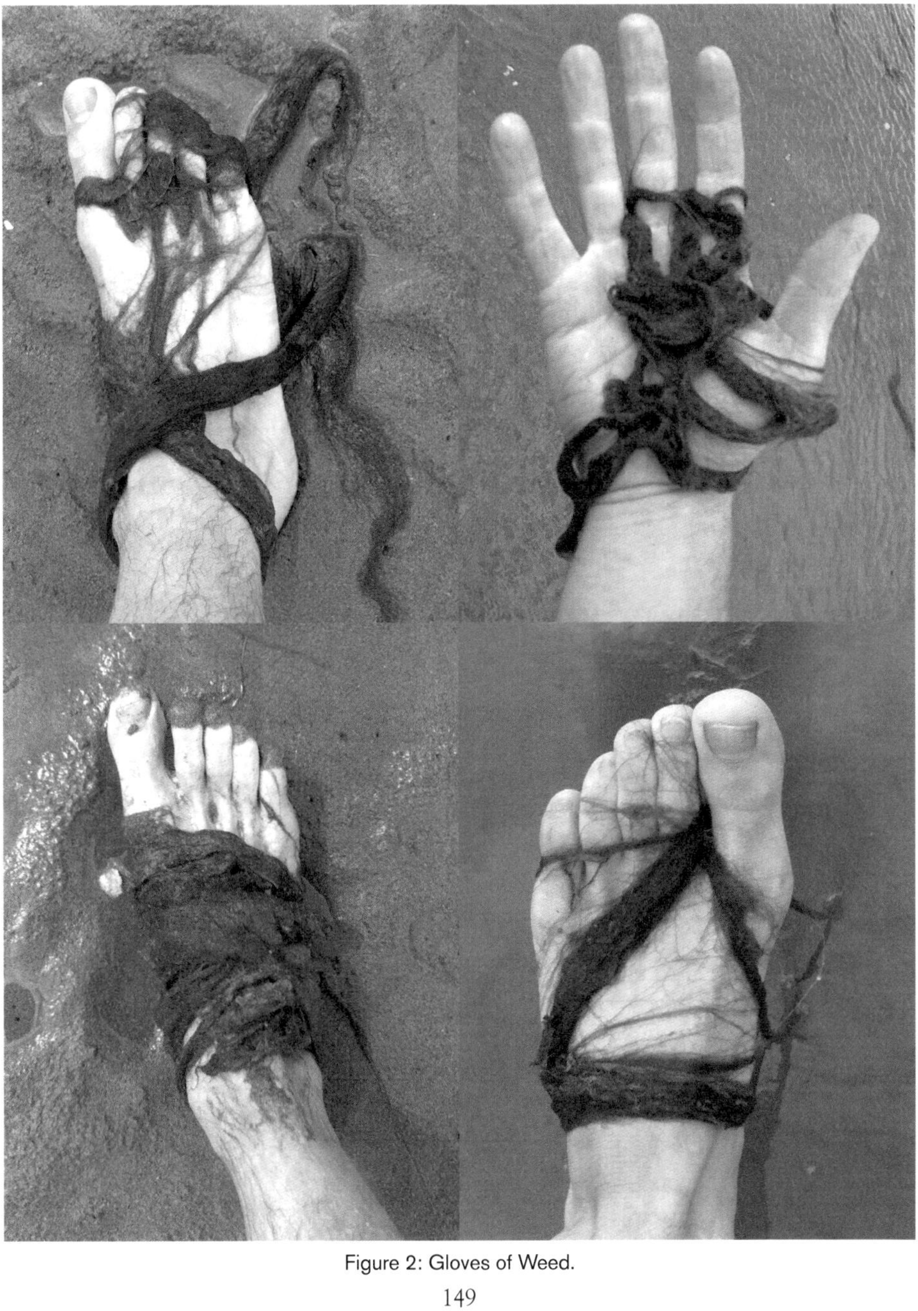

Figure 2: Gloves of Weed.

As a child I feared the feeling of seaweed wrapping around my legs when I swam in the ocean. While Earthbond Prototyping I contemplated this story of discomfort while walking through the shallows of a sandy beach. I picked up algae and seaweed and threw it in the air, tossed it from hand to hand and wrapped it through my fingers. The simple intuitive motions of my body created weed configurations. I was building structures on my hand, creating gloves of weed (fig. 2). This escalated into swirling my feet in the tide pools for a moment; shoes of weed appeared. I repeated this thirty times getting a unique sandal form with each swish through the brine. The seeming randomness of the seaweed entwined with my motions supplied the changes necessary to quickly generate iterations. In a short time the discomfort of being tickled by the seaweed disappeared. I had built a connection and wondered how I could get more people to walk through the shallows barefoot so they too could design with nature's influence. (Camozzi, 2018: 35)

Embracing and acting on the unexpected outcomes from making with self originating materials and phenomena are learning opportunities that develop ideas and content for creativity. Making in this way is difficult. It takes much more time than you would expect without the tools, assumptions and systems abundant in a studio and often there is an audience interested in the process that can be temporary collaborators or off putting. Even collecting enough materials to work within a human scale can be a challenge that has large ecological impacts. Try making small things for a small audience, non-human or inanimate. Each exploration builds positive and negative relationships between artifact and environment that you can observe. My path of consuming the algae exposed light to the water column. This potentially gives more beneficial species the opportunity to compete in the delicate ecosystem. I even remembered that algae thrive in phosphorus and nitrogen rich water and began to wonder if run off from the surrounding homes was causing the bloom.

5 *Attend to nature*

Over time your surroundings will teach you what is "appropriate to do there" (Chambers, 2008: 120). Seek an understanding of the cost of design actions to ecology. Direct your view and reflect on your intervention. What imbalances and opportunities did you create? Are birds feasting on the worms you exposed when digging? (fig. 3). If you return in a week is your leaf display the home of countless insects? Capture these events through autobiographical writing that includes senses, feelings, thoughts and emotions. This heuristic technique and dissemination tool shares how you are changing and gives you an opportunity to grapple with why a radical transformation of the environment in pursuit of designs at this scale seems appropriate for the outcomes. Ask yourself how far does your impact reach? While Earthbonding you may severely disrupt a micro-ecosystem but you are unlikely to cause lasting impacts. Even more interesting to me is this is mitigated with the awareness this method brings. Conversely, acts of making in a studio separate us from consequences. Material extractions, energy for refinement, transportation are often too abstract or vague to visualise or quantify. Furthermore, the convenience at which we can replace used resources allows us to take large risks with materials. When we neglect the flux in relationships between artifact and environment in rich lived contexts, we are no longer designers.

Figure 3: Sand Castle Relationships.

6 Recognise biophilia in your design and its influence on designed outcomes and others

After all of these steps you might be tempted to recreate the dappling of light that shines through the shifting leaves on a tree. This Biomimicry is a design avenue to follow but one that doesn't invite others to spend time under the flitting canopy of a local tree. Connecting to nature, the thing we feel but often can't explain, is complex, non-linear and full of the unexpected. It leads to many more discoveries, a love for the world around us, stewardship and action. An object can mediate this but not supply it (Verbeek, 2005), so look hard at your ideas and outcomes and determine if they support further connection. Help your audience cross their own thresholds to intertwine design with nature to discover nature's intrinsic value and contributions to wisdom, creativity, and design.

Earthbond Prototyping as a strategy has helped me and others be present in new environments, letting research and design opportunities emerge from simply taking the time to loosen control, reflect on relationships and then act. I now apply this method within my conventional design process, rough prototypes are brought into context as early as possible acting as a catalyst to discovering the control to loosen. It takes time but often exposes relationships I had not previously considered.

Earthbond Prototyping is also incorporated into an interdisciplinary design field course at Emily Carr titled This Forest Thinks. It gives students the opportunity to practice deep observation in nature through weekly field trips to local parks. Students are encouraged to find their own approach to conversations on design and nature through practice.

REFERENCES

— Bai, H. (2012). Reclaiming Our Moral Agency Through Healing: A Call to Moral, Social, Environmental Activists. *Journal of Moral Education*, *41*(3), 311–327. Available at: https://doi.org/10.1080/03057240.2012.691628

— Camozzi, Z. (2018). *When Nature Hacks Design*. Thesis (MDes), Emily Carr University of Art and Design, Vancouver.

— Chambers, C. (2008). Where Are We? Finding Common Ground in a Curriculum of Place. *Journal of the Canadian Association for Curriculum Studies*, 6(2). Available at: http://jcacs.journals.yorku.ca/index.php/jcacs/article/view/18099

— Classen, C. (1999). Other Ways to Wisdom: Learning Through the Senses across Cultures. *International Review of Education – Internationale Zeitschrift für Erziehungswissenschaft – Revue Internationale de l'Education*, 45(3/4), 269–280.

— Orr, D.W. (2002). *The Nature of Design: Ecology, Culture and Human Intervention*. Oxford: Oxford University Press.

— Van Der Ryn, S. & Cowan, S. (1996). *Ecological Design*. Washington, DC: Island Press.

— Verbeek, P.P. (2005). *What Things Do: Philosophical Reflections on Technology, Agency, and Design*. University Park, PA: Penn State University Press.

Timo Rissanen, Laura Sansone

Design Students in Sustainable Systems

This chapter outlines pedagogies that we have implemented in the Sustainable Systems course at Parsons School of Design in New York City. We highlight resourceful experiential learning in a hyper-urban setting, where connecting with nature within a three-hour class can be a challenge. Sustainable Systems is a required course for over 1,000 first-year undergraduate students across 11 disciplines at Parsons. We have both been involved in the course since its inception. The course introduces students to the systems inherent in adaptive, resilient design practices. The curriculum is structured around climate change, which frames students' observations as they study real conditions that are both local and accessible. Both fieldwork and applied research methods are developed into creative works that support diversity, adaptability and resilience in the face of our continually changing climate. The *Okala Practitioner* (White et al., 2013) is a required text in the course, providing a shared language across a diverse student and faculty body. The pedagogical design of the course reflects nature's diversity: the course does not impose a specific approach on teachers to deliver on the course learning outcomes. As teachers of the course we both employ experiential, project-based learning, drawing from our personal experiences with nature's systems. Our approaches highlight small-scale production systems in which interdependencies between nature and communities are inherent.

Timo Rissanen: Guardians of Life

I grew up in a country town in Finland, surrounded by woods, lakes and farms. At around age six, without telling my parents, I took a potato from the pantry and planted it in the woods. This foray into guerrilla gardening arose from having been told that a potato in the ground would turn into many; this had left me mesmerised and inspired to act. Unfortunately, I promptly forgot the location and missed out on a potential spud boom; however, the magic of possibility in the moment of planting has remained for the decades since. Some years later, discovering dad's birdwatching notebooks from the 1950s inspired my lifelong passion for birds. Along with a systematic attempt at understanding the world, he meticulously recorded both the Finnish and scientific names of all the species he observed, noting any unusual behaviours. In turn, I cannot look at a bird without trying to identify its genus and species. My pedagogy aims to get design students curious about biology and their ability to grow things, and to care for living things.

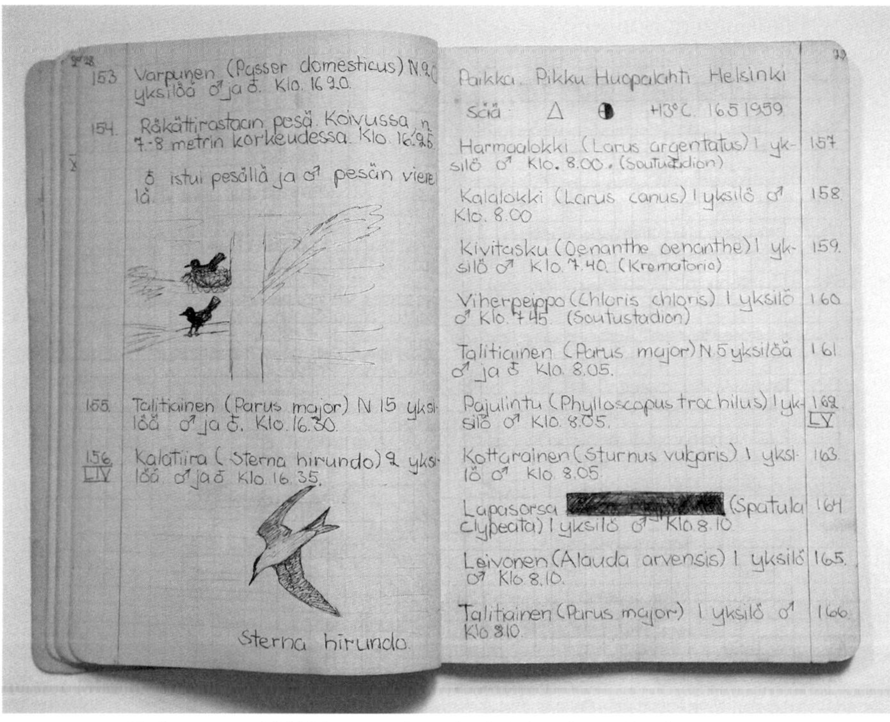

Birding notes of Olli Rissanen, the author's father, from when he was 13.

In teaching Sustainable Systems, I have run two iterations of a project, Guardians of Life. The underlying aim is to foster a sense of awe about life itself while observing seemingly simple biological processes. In the first, each student receives a small succulent plant at the start of the semester. The assignment is to identify the plant and keep it alive until the final week. Students document photographically the growth—or death—of the plant over 15 weeks and compile the photographs into a film. The identification method introduces students to an aspect of the scientific process. Even if a student asks their parent or a flower shop owner what the species is (both of which happened), they must then verify the information. This process led to a rich discussion about the reliability of information. For example, is Wikipedia a reliable source and why or why not? Where one species received multiple identifications, we discussed processes for detecting a scientific error. Approximately half of the plants survived the semester. A year later a student emailed unprompted with a photograph of her plant that had clearly flourished. While potted plants may be mundane as examples of non-human life, they offer a powerful entry into discussing humanity's potential in being guardians of all life on the planet and stewards of intentional climate mitigation.

The second iteration of the project introduces students to the Flower District of Manhattan, a block of flower shops at West 28th Street. At the closing time each day, the stores throw out hundreds of plants: anything deemed visually undesirable and thus unlikely to sell is tossed. A loose community of people emerges each night to rescue these plants, feeding off an existing economy and creating an informal waste management system. This community becomes part of a web of alternative economies in the city, which we discuss in the course. Thackara (2015: 49–66) provides a helpful frame for the discussion of intersections of nature's systems in cities. In the first week, each student rescues a plant and identifies the species, with the task of then keeping it alive for 15 weeks. Students bring the plants for discussion in the second week, and again in the final week of the semester. (By then, some plants have died.) The assignment invites students to reflect on our ability to care for each other and care for other species at a time when we may be the only one (but perhaps not) aware of the on-going extinction crisis.

The unpredictability of the rescued species was more challenging to manage than with pre-assigned succulents, while making connections to communities and economic systems was easier in this version. One student's inability to distinguish between cut flowers and a potted plant with a viable root system became an opportunity, albeit without preparation, to discuss plants' ability or not to develop new root systems from cuttings. Regardless of location, design educators have countless opportunities to connect pedagogy with regional economies and local communities; this example highlights a local, living waste stream around which a thriving community has developed.

Laura Sansone: Urban Dyer's Almanac

I left NYC almost two decades ago to live in the Hudson Valley, a geographic area comprised of small farming and industrial towns residing in the basin carved out by the Hudson River. Living adjacent to the river among agricultural communities makes me keenly aware of the rhythms in nature, like the sowing and harvesting times of plants and the shearing seasons for sheep. Rural life in the valley shifted my thinking about materials and processes that are embedded in the textile design and informed my world-view about economic systems.

For a while, I lived and developed textiles on a fifty-acre farm that was cultivated by my neighbour, farmer Joe O'Brien from Healthway Farms. Joe grew apples and vegetables that he carted down to sell at the Union Square Greenmarket in the city. On the farm, there was an on-going fascination about the growing cycles of plants as I watched each season come and go. Nature has a way of self-regulating and diversifying to create balanced ecosystems. There were never any materials wasted on the small-scale farm. My directive was to design textile products that did the same, and I planted gardens to grow natural dyes, built compost piles to break down textile scraps and used the wool from community farms to create a healthy textile design-ecosystem of sorts. I even enlisted Joe to collect his onion skins for me to use as natural dyes before they landed in the compost pile.

I travel back to NYC regularly to teach textile design courses and coordinate the Sustainable Systems course at Parsons. The weekly travels along the Hudson River draw more than just a geographic connection for me between the urban metropolis and rural living. There is an economic link between the two regions, as many farmers like Joe O'Brien travel to the city to sell their locally grown products at the urban greenmarkets. Just as the regional farm cultivates a varied ecosystem, the local market fosters economic diversity, providing a venue that helps to grow small production systems. It is a network of small, interconnected businesses and services, including vegetable growers, flower producers, meat and fibre farms, as well as compost and used textile collectors. Plant and material life cycles inherent to small farming operations are accessible in the urban context of the greenmarket and it has become my teaching laboratory for natural dye exploration, regional fibre sourcing and textile recycling. Concepts around localism and 'living economies' are outlined in the writing of Vandana Shiva (2005: 64–71).

The curriculum that I teach at Parsons examines the potential to develop natural dyes from the plants that are integral to our regional food economy. In our textile studies, we use the by-products of the harvest, for example, onionskins and carrot tops from the vegetable growers, wool from the meat farms and beeswax from the honey producers. We apply these materials to develop textiles through natural dyeing, construction and surface design techniques. As design students extract colour from plants their data are collected into a compendium of dye recipes called

The Urban Dyer's Almanac (http://www.urbandyersalmanac.com). The UDA was first developed in 2010, and we continue to add to the data of dye recipes today. Our database of dye experiments is organised seasonally according to the time of the year when the plants are locally accessible. This design research suggests a textile design ecosystem where products follow the seasons of the harvest, planting and animal shearing, rather than being directed by market trends that are reliant on profit demands alone. As designers link to the economy and environment of the regional farm, they discover an intimacy with natural processes and materials. The result is new and broadened understanding of the relationship that exists between design, nature, material production and small economic systems. This exploration gives students insight into how they might design adaptive solutions to climate change through the lens of nature and community.

Students developing natural dyes and textiles at a local market in New York City.

Nature as a wayfinder in design education

In both of our teaching approaches, we address the waste streams of existing economies in order to give rise to new economic systems and to build communities. People rescue plants from the Flower District for their own uses, and alternative textile dyeing methods emerge from the by-products of the food and flower growers at the farmers market. These activities are accomplished through shared experiences between students, faculty and the public. Both of these examples highlight various interdependencies between communities and nature's systems. We both employ experiential learning to inform the connections between locality, nature, design and economics. Project-based research enables our students to study how design can act as an agent of change for both new and existing systems. We ask our students to consider: what would happen if design and production were synchronised with the cadence of nature? We believe this would foster homeostasis within systems and work toward building interdependencies. For example, the scavenged plants could become the source for dyeing, and be included in the Urban Dyer's Almanac database, and subsequently integrated into new product development. For us, the natural world has shifted our thinking about scale, time and economics, and it shapes how we teach design. It is the responsibility of art and design educators to teach the next generation of designers and artists to be adaptive and resilient in the face of our volatile climate conditions, and the economic and social instability those conditions create.

Preparing sunflowers to make natural dye.

These project-based activities result in behavioural shifts in students as both producers and consumers. For example, after graduation some students are developing new natural dyeing services, while others are incorporating biological processes into clothing production and art. Our methods build new relationships among teachers and students fostering a sense of community through exploring and creating with nature. In Rissanen's class students share information about succulent care in the class throughout the semester: guarding the life of the plants becomes a team effort. Natural dyeing, while an entry into teaching about nature's systems in relation to textiles, is a community-building activity as participants share dye research. Sansone runs a series of natural dyeing workshops to different classes; this networking is important in maintaining a sense of community across 32 simultaneous sections of the class. At the end of the semester, all teachers and students come together in a studio crawl to celebrate the communal learning. We are reminded that using natural systems as a model can help us conserve our precious natural resources. In a hyper-urban campus, the Sustainable Systems class fosters a crucial connection with nature among the students.

REFERENCES

— Shiva, V. (2005). *Earth Democracy: Justice, Sustainability and Peace.* Cambridge, MA: South End Press.

— Thackara, J. (2015). *How to Thrive in the Next Economy. Designing Tomorrow's World Today.* New York: Thames & Hudson.

— White, P., St. Pierre, L. & Belletire, S. (2013). *Okala Practitioner. Integrating Ecological Design.* Phoenix, AZ: Okala Team.

(11 December 2017) Snow still here and today bright skies. I skipped and slithered in the hills this morning, smiling all the while.

Cameron Cartiere, Nancy Holmes

Design on the Wing: Collaborative Work with Nature

We're two artist academics in British Columbia working with bees—native bees, for the most part, not honey bees. We're collaborating with bees around their habitats: designing meadows, pastures and flowering hot spots. Neither of us has had a significant history with bees, though we both garden and love the natural world. However, we became inspired to do something for bees when Cameron attended a talk by Dr. Elizabeth Elle. She is one of Canada's foremost pollination experts and she was eloquent about the plight of native bees. She said that scientists around the globe were helping honey bees deal with diseases, colony collapse disorder, mites and pesticides, but very few people were aware that native bees were also experiencing grave threats. She said that the best thing non-scientists could do was make and protect habitat. She also wished that someone could reveal the wonders of all the countless kinds of bees around us. Most people think there are only two types of bees: honey bees and bumble bees, and the latter they're not sure about. Yet in British Columbia alone there are over 400 species of bees, over 900 in Canada and over 20,000 worldwide. Who knew?

So, with Dr. Elle's blessing, we took up the challenge. We were makers, Cameron with things and Nancy with words. We would talk bee, show bee and make bee habitat using all we know about art, metaphor, design and community engagement. We had a lot to learn, but bees turned out to be great teachers. They are exceptional makers and dazzlingly diverse. We began to design and create pollinator pastures and other unique habitat projects in our two communities, with the first lesson about collaborating with nature being: *open your eyes*.

see how bees and flowers
designed each other
see how bees design their homes
yes, hexagons of honeycomb
but wild bees too
architects of mud walls
plunging tunnels, leafy barriers
resinous bowls and pots
see the little master builders
sabotaged
by the anthropocene

From the beginning, our methodologies of collaboration needed to be adapted to work with two difficult communities: 1) nature in the form of damaged land we were going to transform into bee-rich habitat; and 2) people in the form of municipal departments and officials. With both collaborators we needed to adapt our artistic and community-based methodologies, consider new approaches and adjust to unforeseen results. Our process involved constantly going back to the aims of the project to remain anchored in the true motivation for the work, motivations both environmental and aesthetic. If the spring was so wet we couldn't plant any seed, we remembered our goal was to let nature lead. If the municipal official thought the meadow should be mowed because it was too messy, we reminded ourselves that we knew it would be an uphill battle to change a sense of aesthetic. This return to our intentions helped us stay open to unexpected opportunities. When spring is wet, we learn about the soil's capacities and textures. When we talk to an official about how many more bees appear when weeds are left to bloom, parks people become a bit uncertain. Practising methodologies of letting nature lead and, then, leading—or nudging—rigid institutional structures, we found collaborating with nature was about listening and talking, as well as looking.

Nevertheless, we were soon confronted by how difficult it was to listen to nature. We looked at the "lay of the land" and in Richmond's long, narrow dimensions (1,000 running feet under the flight path to Vancouver International Airport), we saw flow and connectivity. We saw a human audience that could move through the earthwork and also

Figure 1. Richmond pasture seeding (2016). Credit: Geoff Campbell/Border Free Bees

view it from an airplane above. We envisioned a planting pattern of outstretched bumble bee wings in continuous bloom from early spring to late fall. Conversely, with Kelowna's curved creek, leaning trees and furrowed cliff-face, we imagined a sense of home. Within that enclosed space we saw circles of native flowers (circles reminiscent of bumble bee nests) offering up opportunities for the temporal and historical context of the site to filter through up through the soil. Yet in each instance our designs were blurred, erased and grown over by nature. Our privileging of how we wanted humans to see the land was simply that—a human-centric view of the pasture and belief in the power of our designs. We forgot the powerful forces of plants and what the bees did by their pollinating. Before we knew it, the design was not ours but nature's. The bumble bee wings became a tangled tapestry; the bumble bee circles were scribbled over by invasive bindweed.

So we learned: always, nature speaks. Design that listens to as well as looks at nature requires a form of slow design. Nature speaks in code, signs and signals. You have to know what to watch out for, to be alert to changes in mood and tone. It is a subtle conversation at times, and at times a full force confrontation. For weeks, nature would give us hints that we ignored and then one day, *boom*: a full infestation of bindweed or a withering drought after a spring of flooding. Seemingly out of the blue, but really, coming on for quite a while. This is a different way of communicating. It takes time and curiosity—looking, hearing and feeling.

Getting your hands dirty—literally. We learned we were working even when we were just standing there, if we were taking note of what we saw, heard, smelled, felt.

the ghostly presence of before
of absent bees, absent
flowers, absent bacteria
absent sound of wind in wild grasses
dreaming design through
veils of loss
certain, so certain,
we can build a semblance
of before

Figure 2. Kelowna pasture in process (2016).
Credit: Fionncara MacEoin/Border Free Bees

Since we worked through our senses and a heightened attentiveness, the pollinator pastures surely were artistic productions. Yet, our pastures taught us about a design conundrum. We began with faith in human design capabilities, believing in techno-scientific progress. We thought it a fairly simple matter to transform an empty stretch of lawn into a flowery pasture. However, we discovered that nature's design is far too complex for us to even begin to understand or emulate. Even proceeding with caution to shift the ecology of a small patch of land has huge risks of cascading unintended effects and inherent failures. We learned to be suspicious of the human intention to manage nature or design it into health and beauty. Yet we value the urge to restore injured or neglected ecological spaces. This longing to repair something damaged is a key imaginary of our time.

We thought we were imaginative and sensitive restorers: our repair of this damaged land would be tempered by past use. Often, though, it turned out we were paying attention to mostly human traces. We noted the desire paths in the pastures created by human users and these paths influenced both the Kelowna and Richmond designs. We knew it was proper community-based design to respect historical use. But communities are also difficult collaborators. Graffiti appeared on the benches and on the native bee apiaries we constructed in Richmond. In Kelowna, homeless campers trudged through the pasture right through our experimental test patches. These human marks were then criss-crossed by more-than-human effects. A marmot den emerged in the middle of the Kelowna pasture and deer paths appeared, meandering through our newly planted and soon-to-be-eaten seedlings. Off-leash dogs ran through the Richmond pasture, spreading trails of invasive seeds, their passages mapped out a few short weeks later as the horsetail plants sprang to life. Like the unexpected complaints of neighbours or the intransigent landscaping regimens of Parks departments, the changing and unexpected more-than-human agencies created feedback loops from which we needed to reassess, adapt and accommodate as new voices along the spectrum of human and more-than-human asserted themselves into the conversation.

feet, tires
twinning prints of deer
claw marks, scars
print over paw
go, come, trample
drag traces
on top of other traces
the bees lift off
the plants lie down

As we learned to read our pastures, we became schooled in the idea that place is not just a surface on a map wherein external realities are to be "solved". Doreen Massey, the feminist social geographer, positions place as "a simultaneity of stories-so-far" (Massey, 2005: 130). As artists, we couldn't help creating a metaphor-laced design that layered stories, readings and images. At the heart of the two pasture projects were scarred, abandoned or damaged landscapes in urban industrial zones. They were plastered over with concrete and warehouses, hydropower lines and gas pipelines. They were so stripped of natural functions that their former "original" qualities were nearly extinguished. Industrial zones are a common urban design feature, the sculpting of spaces devoted solely to human activity. We learned that the very placement of bee pastures in the middle of such neglected industrial zones was a design feature that decentred the human. How often those of us working on the pasture were told we were creating a "hive of activity", that we were "busy as bees" and "buzzing with creativity". The pastures reframed the idea of "busy" or "industrious" to assert the industry of more-than-human beings. The geographical situation of these two pastures made the claim that when land seems to be "inactive"—no development or human use—it doesn't mean that the land isn't "in use." Such spaces are infiltrated by the subversive natural unseen: crows occupying wires and weeds erupting through cracks. And similarly, our pastures disrupted the very idea of an industrial zone by highlighting the industry of other species, an industry that interacts with the land, that does not ignore it or "solve" it. This was a collaboration with nature where art worked as subversively as nature.

in solvent-polluted air
hands feet wings
store remembered work
yellowish powder in the wind
delicate apocalypses
on zinging petals
busy thread-like feet
fabricate, haul
loads of future
into the here
and now

In the end there is the great feedback loop of death, always death, but also rebirth and regeneration. This cycle is Nature's primary metaphor and essential design. Nature is brutal but also forgiving. Sometimes it's rain, sun, heat, rain—all at the right time and all in the right order. Then again, it can be rain, rain and more rain or endless days of unrelenting sun. In Richmond the pasture flooded, and trees fell down. In Kelowna a summer of three months of drought fried every new seedling. The challenge is to know what to do when the tables turn and how to work with what the land is telling us it can give, not just what it needs. Perhaps the most valuable outcome was the understanding that collaborating with nature requires the ability to take time. Time to understand a specific site in various conditions, under a range of influences and over the cycle of complete seasons.

We are often so focused on instant gratification, on immediate results. We get a grant: create the project, provide the outcome and evaluation one year later. But how can a single year measure success for projects that are planned for decades? Projects that are meant to confront years of abuse, neglect or poor planning? How can we design for death, for rebirth, for reflection? Over the course of time, how does nature measure success? Our bee count and diversity are up—is that a success? It is a success for our science partners—but is it an artistic success? And will it last and is it enough?

From Rachel Carson's prophetic vision of a "silent spring" in 1962 to George Monbiot's "insectageddon" in 2017, we know we are the guilty ones who have catastrophically designed a world that has erased millions of beings from hundreds of thousands of gutted landscapes and bodies of water. So, like countless other well-meaning people, we embarked on projects of habitat creation—to work with nature, to collaborate, to transform neglected patches of land into places that could support pollinating insects. We began this undertaking, a bit blind, a bit deaf, a bit complacent. We walked onto the sites whose stewardship we had accepted and encountered the spectres of the past, the dreams of promise, the chimeras of our own thinking with its desire paths and dead-ends.

REFERENCES

- Carson, R. (1962[2002]). *Silent Spring*. New York: Houghton Mifflin Harcourt.
- Massey, D. (2005). *For Space*. London: Sage.
- Monbiot, G. (2017). Insectageddon: Farming is More Catastrophic Than Climate Breakdown. *The Guardian* [online] 20 October. Available at: https://www.theguardian.com/commentisfree/2017/oct/20/insectageddon-farming-catastrophe-climate-breakdown-insect-populations

Clara Vankerschaver

Sustained Observation and Processes of Growth

Growth concerns us. As humans we grow crops and raise animals, activities that constitute the domain of agriculture. We also grow ourselves, sometimes to our disliking as in *growing old*. Moreover, we have become accustomed to using the term referring to economic growth. Essentially, growth is a process in which certain conditions allow animate things to develop and possibly prosper. However, growth also characterises the way in which *inanimate* things, objects, come into being and thereby reframes the field of production. Anthropologist Tim Ingold suggests that objects "emerge—like the forms of living beings—within the relational contexts of the mutual involvement of people and their environment" (2011: 88). This offers a different lens through which to consider artisanship and design practice. If these domains are similarly shaped by processes of growth, how does that affect the role of the human agent involved? If all things take shape in a field of relationships, how does this actually come into play in the making of objects?

To further explore this line of thought, I bring into account my own practice as a textile designer. Textiles, as tangible objects and surfaces surrounding us, are familiar yet at the same time they are more obscure than ever. Paradoxically, textiles, ubiquitous as they are, remain largely unnoticed. We often fail to devote special attention and while there generally seems to be an increased awareness about ecological and social issues, transparency about provenance and production is still greatly lacking. A deep-rooted fascination with textile materials led me to trace down origins and people involved in the very first stages of production. Fascinated by ways of working that evolved over time my attention mainly turned to natural materials with a long history of use. Wanting to know from the inside, I ended up growing my own crops and thus becoming, in a way, a farmer myself. This experiment started three years ago and has since evolved to be an integral part of my practice. With an initial focus on growing cereals to harvest the straw as an experimental textile material, I prepared land and nurtured plants following natural cycles that have their own rhythm. The challenges were manifold: finding a large enough stretch of land, accepting that not my schedule but seasons and changes in the weather allowed specific work to be done, adjusting to the relative large scale of my field and finding tools accordingly … . Responding to these challenges was both profoundly humbling and enriching. Even if my plot was only a fraction of what an actual farmer handles, it was large enough to raise an acute awareness about the interdependency of natural systems and why this is such a fertile perspective.

Learning by looking, I spent much time out in the field, seemingly not doing much. That time spent looking cultivated a tacit understanding. Farming makes you explicitly aware that crops do not grow because you want them to but because you help to create the conditions in which they thrive. Nurturing means noticing what those conditions are and how you might constructively influence them. Simply standing by the side of the field, for example, I could feel the wind and simultaneously see how it moved the oats that I remembered planting at a hand's width. While noticing this I wondered how the density of the crop might help it to either stay upright or be blown over and eventually wither. And what would happen if it rained heavily? I was familiar with the general guidelines and predictions, but which manual could have told me of the weather conditions in this particular area and time? By attentively observing, I was able to develop experience that benefitted next season's crop. Although my general knowledge and experience as a farmer remained very limited, it appeared I was "qualified by commitment" (Marchand, 2016: 9). Attuned to my specific circumstances I was able to intervene when needed and yield a qualitative harvest.

The incentive to gather and grow my own materials had been to know more from within and less from without. Yet even if these experiences were very specific to certain crops and ways of farming, their meaning can shift domains. For me, it required the firsthand but unfamiliar experience of nurturing plants to fully realise that in making objects the same processes apply.

Techniques for the production of textiles have not changed that much since they were first developed. The same basic principles of threads rhythmically crossing or looping still make the specific structure of fabrics. However, in the current industrial system these plain procedures are made extremely complex by unprecedented expansion and economic imperatives. Modern industrial logic imposes the time of economical imperatives to the rhythms of ecological processes (Mathevet, 2012: 162), eroding the ecological capacity for renewal. While a certain large-scale industry might be needed to provide products democratically, we should continuously question its preconditions. Considering growth as a significant process in the making of objects makes us think about the conditions and timescales that apply to manufacturing processes. Profit-driven contexts are increasingly undermining the intrinsic logic of genuine growth, which works with different and cyclical timescales and innovates by gradual adaptation. Working from the perspective of growth means doing things

Reed harvest in the Camargue. Reed is harvested during winter and spring, before new shoots appear. Riding on the back of an amphibious machine, two or three mainly casual workers alternatively load and stack the sheaves of reed, all day long.

over in response to the shifting contexts of life. By acknowledging the influence of the circumstances that brought an object into being we are prompted to step into different and often highly specialised fields and assert the importance of access. While the network society heralded open access, enclosure is progressing in unforeseen ways. Access to diverse kinds of knowledge—often our cultural heritage—and access to land, seeds, materials, tools, is frequently impeded not just by physical borders but also by law, pesticides, privatisation etcetera. Working with farmers-manufacturers in the Camargue region (France) I was recently asked to redesign products made with reed, seasonally harvested in the local marshes. After getting acquainted with their ways of working I had difficulty finding what exactly needed to be redesigned about the end product. Following the logic of working in the marshes the final objects produced made perfect sense. However, difficulties arose because specific rights to harvest reeds have been lost in recent years and formerly common land has been increasingly privatised. A close—even bodily—connection to the marshes, its ecosystem and social history attested to a sense of stewardship, which from a mere economical point of view seems otherworldly.

> *You let the cut straw bathe in a shallow pail filled with water and weigh it down with whatever is close at hand. Oblong shapes—scissors and sticks this time—keep it submerged until the water gradually finds a way inside and the straw becomes soaked. You figure that the material needs some time to adapt to its new state and you let it bathe a while longer. A laden sound tells you when it can be taken out. Running a stem between your thumb and index finger you squeeze the excess water out and witness how the material has become pliant once more.*

Moving beyond the classical dichotomy between animate and inanimate matter *cultivation* takes on a broader meaning. Inanimate matter too has a will of its own and accordingly allows or rejects certain manipulations. There is a saying amongst fishermen that when making fishing nets the thread has to *fight out its anger* before it can be neatly wound onto the netting needle to make a net that does not tangle. Cultivation means to develop and improve by being aware. It will only surprise a complete novice that a deliberate design is not simply "written out in the material" (Ingold, 2011: 343). Maker and matter engage in a dialectical process of give-and-take to arrive at a new form. Becoming skilled is a social process and means getting acquainted with the material and tools at hand. Still, failure always remains a possibility. While failure can be frustrating, it can be a beneficial experience. The numerous cuts in my hands while processing cereal straw are a testimony to a steep learning curve. With each cut my attention was explicitly drawn to a point of contact between the material, the tool and myself. I patiently endured these often-painful mistakes that reminded me to work *with* instead of *against* the material and tool. It is, in short, how I learned to wield a sickle. Clearly, it takes an unassuming attitude to listen to what materials and tools are able to tell you, to what they afford. Anthropologist Trevor Marchand confirms that "tools have a history of use and each kind has been designed and manufactured in a way that suggests how to grasp and use it" (2016: 17). These affordances are always also influenced by the overall context of the work. While encyclopedias and their contemporary variants may disclose many processes otherwise hidden in workshops and manufactories, they never give a complete picture. In actual practice, genuine insight relies on sentient engagement with materials, tools and context. A strictly technical reduction will always fail to tell all that needs to be known.

Embodied knowledge pervades practice but is not easily made explicit. Often belittled or mistrusted, these sentient and tacit ways of knowing have been banned from the canon. Finding appropriate forms to tell about these insights is needed to reframe the dominant discourse that aims to make everything quantifiable. Although it is meaningful to use diverse media, thoughtfully using the spoken and written word remains a way to open (individual) experience and make it last beyond the moment. Noticing without being able to name specific features of the landscape or certain phenomena their context was lost to me. Local dialect and idioms, often featuring in the language of seemingly *uneducated* people and even children, are a great source of knowledge about the landscape as author John Stilgoe reminds us (2015: 72). Though language helps to communicate it also serves in sustaining observation. We notice more when language helps to diversify what we notice. Light, color and atmospheric conditions are frequently well known by locals, yet often unseen by schooled inquirers because standard dictionaries frequently fail to include vernacular vocabulary (Stilgoe, 2015: 72). As in references to specific landscapes, many crafts have their own unique lexicon with synonyms hinting at significant variations. Similarly, etymology helps to reveal hidden relations and the historical development of meaning. In my recent practice I came across the word *paillasse*, which is used in both French and English meaning a straw matress. In French the word for straw is *paille*, indeed indicating that a *paillasse* and other variations such as *paillasson* are made of straw. Interestingly, in Dutch *paljas* is used to name a specific kind of fool. This other meaning can be traced back to historical plays where a living scarecrow with straw sticking out of his clothes is staged as the ultimate buffoon. Here straw, a common and scrap material, becomes a reference to the lowly background of this character. This brief excursion into etymology already reveals some aspects of the historical and social context of the use of this specific material.

The call of actuality to the designer is to reconsider what it means to make things. It is challenging in a society that is increasingly built on standardised procedures and quantification to develop a dialectical practice, which suspends the desire for absolute closure and responds to

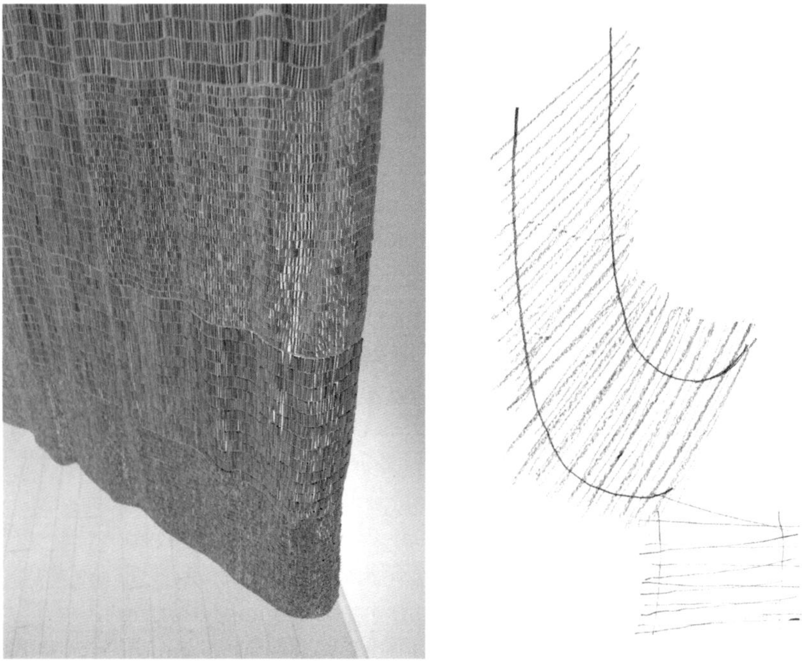

Samples and sketches. In getting acquainted with specific materials, sketching and sample making are essential. The rhythm of craftwork as a process of growth is defined by continuously 'going back to the drawing board'.

broader field of attention. With this increased awareness of circumstances, we go beyond our individual practice and we are challenged — but not unable — to see the thing as a whole, the wood for the trees. It is what I have experienced when as a designer I stepped out in a different field. An ancient field that for a contemporary designer yields not merely resources but more importantly insight about ways of working within the natural environment on which we all so greatly depend. Continuously going back to this beginning is what allows us to move forward.

REFERENCES

- Ingold, T. (2011). *The Perception of the Environment*. London: Routledge.
- Marchand, T. (Ed.). (2016). *Craftwork as Problem Solving*. London: Routledge.
- Mathevet, R. (2012). *La solidarité ecologique*. Arles: Actes Sud Editions.
- Stilgoe, J. (2015). *What is Landscape?* Cambridge, MA: The MIT Press.

Maarit Mäkelä

A Nourishing Dialogue with the Material Environment

Encountering creative materiality

Social Anthropologist Tim Ingold (2004: 330) reminds us that we do not perceive with the eyes, the ears or the surface of the skin, but with the whole body. It is specifically through our feet, in contact with the ground, that we are most fundamentally and continually 'in touch' with our surroundings. His argument is that it is both the hands and feet—augmented by tools—that mediate a historical engagement of the human organism with the world around it (ibid.: 332). In Waiheke Island, the place where I lived during my nine-month stay in New Zealand, walking emerged as an embodied practice that underpinned the creative process that took place during the year.

Walking was a multisensory experience during which my body perceived its surroundings through a diversity of senses—seeing, hearing and smelling—in combination with a moving body (Mäkelä, 2016: 5). At the beginning, the purpose of the walks was to familiarise myself with the new environment. I was impressed with the mineral variety of the land that I encountered in the form of rocks, stones, sand and soil—next to roads, walking tracks and beaches, and under the sea when the tide moved the water further from the beach. When I found captivating materials, I handled them to discover more about their qualities. Whenever I would be able to grind these materials with my tools, I gathered up a small amount of them and took them with me.

The first such gathering was the black sand that I found during my walk at Te Henga beach. This was a unique experience as I had never before seen a black beach that was entirely formed of black sand and black stones of varying sizes. The soil was volcanic and contained a lot of manganese that lent its colour to the entire environment. The heavy waves washed the beach, transforming the stones into sand. The rhythmic sound of the waves followed my visit in the beach, accompanied by the smell of the salted sea. During my walks in different locations, I ended up also gathering stones and clay. As a ceramicist, I was eager to discover how I might be able to use these materials as a part of my evolving creative practice.

The political theorist Jane Bennett has considered the relationship between nature, ethics and affect. Recently, she has shifted her focus from the human experience of things to the things themselves and considered what it might mean if we were to recognise the active participation of nonhuman forces in events. Her thinking starts from the conceit that humanity and nonhumanity have always performed an intricate dance with each other (Bennett, 2010: 31). Despite acknowledging the complexity of the dance, the most important agential factor has been positioned as human intentionality. That is, humans have been conceived as being the bearer of an exceptional kind of power (ibid.: 34).

In *Vibrant Matter* (2010: 56), Bennett proposes that, instead of a formative power detachable from matter, a craftsperson—or anyone who has an intimate connection with things—encounters a creative materiality with initial tendencies and propensities, which have a capacity to be combined in varied ways. The direction in which this power takes the creator depends on what other types of forces, affects or matters are present in the process, or bodies with which they come into close contact. What is intrinsic for the craftsperson is the desire to see what the material can *do* (ibid.: 60). Via this position, she is also able to discern a life in the material and thus, eventually, collaborate productively with it.

My approach and thoughts as a craftsperson parallel what Bennett proposes above. Walking in a natural environment invites me to take part in an interplay with it: it gives me both inspiration for the creative work and the concrete materials with which to start the process. In New Zealand, the carefully selected material gatherings formed the material base of the evolving creative process. As I gathered all the materials from different locations, they were all unique—differing from the nature of the materials I was used to working with.

In dialogue with the local environment

Ingold (2000: 88) suggests that artefacts emerge—like the forms of living beings—within the relational contexts of the mutual involvement of people and their environment. He believes that the form of the artefact does not issue from the idea but instead comes into being through the gradual unfolding of the field of forces set up through the active and sensuous engagement of practitioner and material (ibid.: 342). According to his thinking, a skilled craftsperson operates within a field of forces set up through her engagement with the material (ibid.: 347). The work does not merely involve the mechanical application of external force, but calls for care, judgement and dexterity. The action has a narrative quality in the sense that every movement grows rhythmically out of the one before and lays the groundwork for the next.

As a professional craftsperson I endorse Ingold's thoughts and conceive that I truly work in a dialogue with the material forces. To be able to conduct this kind of subtle dialogue with the material, I have to be sensitive to the nonhuman material world that surrounds me, and that I am part of. As cognitive archaeologist Lambros Malafouris (2008: 22–24) has proposed, in this kind of dialogue the question is not about a property or possession of humans or nonhumans but the relational and emergent product of material engagement that has to be realised. What I am able to do is dependent upon the physical affordance of an environment—in my case the qualities of the soil that New Zealand provided to me.

To understand more about the material quality of the soil gatherings, I established a plain studio at my home. The tools consisted of local sticks, shells and stones, but also included some kitchen items, such as towels, spoons, knives and a pestle and mortar. In the studio, I processed the collected materials further with my hands. I refined the materials by separating the small sticks and bigger stones from the gatherings, and sorted them out based on their colour, and the location from which they were collected. After that, with the help of a pestle and mortar, and with some added water, I transformed some of the clay, as well as sand and stone samples, into a liquid form.

I also shaped the rest of the clay and from this clay body made tiny test pieces that were dipped into the liquids I had made. In this way, the same clay served both as a body of the work and as a colour that was applied to other clay bodies' surface. The first test pieces were made from the yellow clay I found on the slope of my own street (fig.). This clay also contained tiny pieces of ochre from which I could prepare a slip that had an orange tone. When the test pieces were dry, they were fired in a kiln at 1060 degrees Celsius. During the firing, the material composition of clay morphed into ceramic, entailing changes in material surfaces and colours. The result of the firings was a tangible palette, a study of the colours the local environment offered to my creative practice. The unfired liquids were used when I proceeded to the next stage of the process and started to paint.

The first paintings were made on paper as I wanted to experiment and see the kind of marking that would be possible with the earthen liquids.

When painting, I let the powered sand, stone and clay run on the paper by using large amounts of water. In this way, I was in a dialogue with the material that 'responded' to my actions unexpectedly. Due to the impure qualities of the materials, some of them rejected each other, and before becoming part of the evolving image, they found their own way to 'dwell' on the surface of the evolving painting. Starting from the abstract forms, the painting process proceeded towards becoming figurative images. After realising that earthen materials were suitable for diverse kinds of paintings, I started to paint directly on top of some wet clay that I had carefully moulded with my hands beforehand.

The change from paper to clay also enabled me to manipulate the painting surface while making the image. Thus, instead of sketching the evolving image with a brush, I used a wooden stick with which I carved the image roughly on top of the wet clay slab. Onto these lines, I added earthen liquids. I worked in layers and the image evolved on the top of the slab in a dialogue with my embodied actions and the earthen materials.

During the painting process, the nature of the lines changed, as I also inserted earthen liquids inside the carved lines and then ran the colours along the lines. When doing this, I took the clay slab together with its wooden board support onto my lap and cradled the emerging painting. Via this action, I was able to lead the running earthen liquids discretely in the desired direction. I consider this part of the process to be very intimate, a kind of caring act that was directed not only towards the evolving image, but also towards the material per se. The clay images were transformed into ceramic via a firing process that eventually brought forth the colours suggested by the test pieces (fig.).

Craft making as philosophical act of care

Geographers Owain Jones and Paul Cloke (2008: 86–87) have brought up the need to move away from treating the human realm as separate, privileged and ontologically unique in terms of agency. To be able to appreciate the agencies of nature and materiality, we need to appreciate the very differing forms of beings and processes in which they are articulated, and, in addition, the very differing velocities and rhythms they might be operating in. Bennett (2010: 108) proposes that we also need to devise "new procedures, technologies, and regimes of perception that enable us to consult nonhumans more closely, or listen and respond more carefully to [non-humans'] outbreaks, objections, testimonies, and propositions". This is because these offerings are profoundly important to the health of the political ecologies to which we belong.

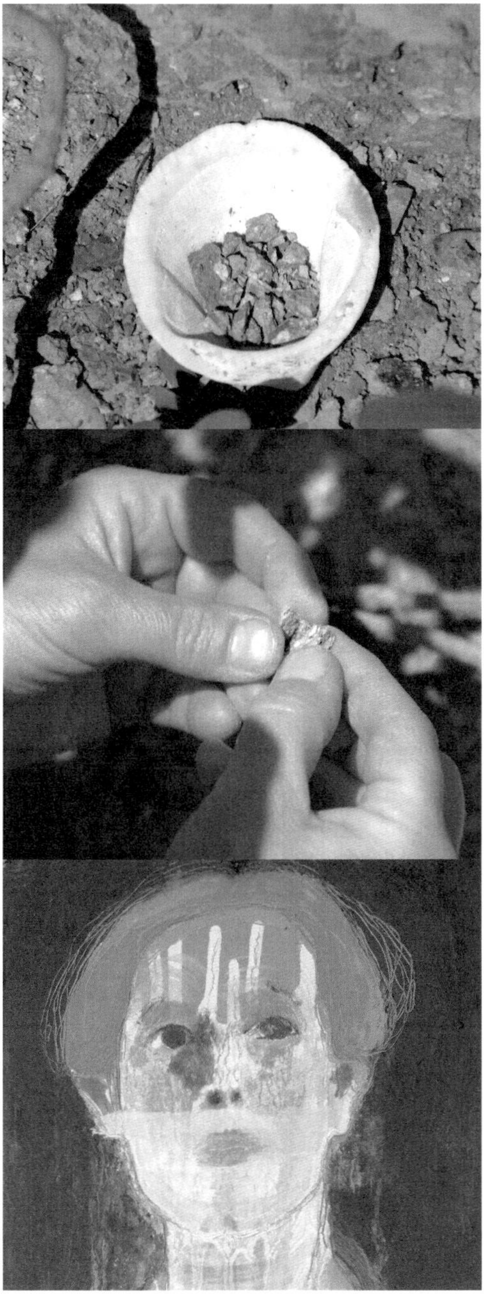

Figure: Collecting earth from Waiheke Island (top and middle) and the final clay painting *Portrait of a Woman* (2015, 26 × 26 × 2 cm) made with local earthen materials: Te Henga black sand, Tasman red stone, Long Bay white and yellow clay, Te Matuku clay and Cory Road clay on Matakana clay. Photos: Maarit Mäkelä (top), Pertti Mäkelä (middle) and Samuel Hartnett (bottom).

In New Zealand, I immersed myself in the local nature and began a dialogue with the new environment. The creative process was set forth via walking, which, according to Ingold (2010: 18), is a deeply meditative practice. He conceives of walking as a journey in the mind as much as on the land. This is enabled through our senses, which maintain a constant traffic between the terrains of the mental and material. During the walks, I immersed myself as part of the local environment and in this way tuned my thinking towards the patterns and rhythms the nature offered. In my case, the surrounding nature not only inspired me and set my mind into a certain mode, but also provided me with materials for the evolving making process.

The evolving creative process was based on my earlier experiences and skills from the field of contemporary ceramics. The process resulted in accidental discoveries and unexpected results that were gathered in an exhibition embodying different agencies and material configurations: small earth gatherings, images from the place where these materials were gathered, tools and test pieces, as well as paintings on paper and clay. Experiencing the interplay of these components in the form of an art exhibition was for me an aesthetic experience.

In this essay, I have conceived of craft making as a philosophical space that enables us to think through the ethical and ecological concerns related to the stage of our environment. Furthermore, craft making provides an embodied way to practise and nourish a dialogical relationship with the local environment. I see this endeavour as a valuable practice through which to (re)consider the relationship between human and nature profoundly—not only from an individual perspective, but also from a cultural and non-anthropocentric one.

REFERENCES

— Bennett, J. (2010). *Vibrant Matter: A Political Ecology of Things*. London: Duke University.

— Ingold, T. (2000). *The Perception of the Environment*. London: Routledge.

— Ingold, T. (2004). Culture on the Ground. The World Perceived Through the Feet. *Journal of Material Culture*, 9(3), 315–340.

— Ingold, T. (2010). Ways of Mind-walking: Reading, Writing, Painting. *Visual Studies*, 25(1), 15–23.

— Jones, O. & Cloke, P. (2008). Non-Human Agencies: Trees in Place and Time. In C. Knappett & L. Malafouris (Eds.), *Material Agency: Towards a Non-Anthropocentric Approach* (pp. 79–94). New York: Springer.

— Malafouris, L. (2008). At the Potter's Wheel: An Argument for Material Agency. In C. Knappett & L. Malafouris (Eds.), *Material Agency: Towards a Non-Anthropocentric Approach* (pp. 19–36). New York: Springer.

— Mäkelä, M. (2016). Personal Exploration: Serendipity and Intentionality as Altering Positions in a Creative Process. *FORMakademisk*, 9(1), Article 2, 1–12.

(11 February 2018) The season is deepening and also there are tiny signs of spring here, I feel everything is running too fast. I don't ever seem to do the most important things, giving space and time to allow them to grow.

Alyssa Schwann

The Poetics of Cultural Landscapes through Ecological Wisdom

A partnership with Indigenous people in the stewardship of the land and in the design process is essential for renewing a shared relationship with nature. My advocacy for establishing these partnerships stems from personal experience in public policy in Canada's North, as well as from research and practice in landscape architecture in Manitoba and British Columbia. This experience of working with Indigenous communities has underscored the necessity for listening to, and learning from, Indigenous Knowledge Keepers, elders, youth and other community members—discussions critical for protecting vulnerable ecosystems and degraded cultural landscapes.[1]

I write as a non-Indigenous Canadian. In writing I do not intend to diminish Indigenous voices, or to act as an interpreter of Indigenous cultural relationships to nature or of the world views of others. What is communicated in sharing circles, or what is gained from walking the land with members of the Indigenous community, must be used neither in haste nor to quench academic curiosity. These are, at root, sacred experiences which offer expert knowledge, knowledge which must be honoured and handled responsibly, rather than appropriated.

There are clear, graspable lessons I have learned from Canadian Indigenous communities. Design must respond to place, to the roots of a location, 'sprung from the land' (LaDuke, 1994); nature is not absent of culture. Indigenous communities are not 'stakeholders' in how we design and develop within our natural systems: they are partners. To evolve collective design practices, we need to ask questions and listen to the answers to deepen our understanding about nature and how to live *with* the land. By establishing meaningful nation-to-nation partnerships we might achieve an integrated study of cultural landscapes and ecological wisdom that informs, or even directs, the process of landscape design and subsequently the conservation and preservation of nature.

Indigenous knowledge of ecology and nature, culturally embedded within Traditional Ecological Knowledge and Wisdom (TEKW), reveals an interconnectedness with land and place. To omit TEKW from the theoretical discourse and practical application of ecology, and by extension from perspectives on beauty and meaning in designing on the land, is to be remiss. I argue that TEKW can simultaneously expand an aesthetic design approach while providing a philosophical framework for the management of natural resources (Nature). Through collaborations that include diverse voices and multiple perspectives on nature and resource management we can expand the possibilities for design. The more we become re-connected to the land—to Nature—the greater our sense of belonging to place, thus restoring meaning to our cultural landscapes. To achieve this position, design must navigate through a new relationship with nature.

Ecological wisdom: a local prospect

Today's design processes consistently omit Indigenous voices. Moreover, in many design discussions the presence of land is often minimised or avoided completely, thereby profoundly undermining Indigenous culture. This relationship of nature and land remains a contentious issue, hampered by the ongoing dismissal of Indigenous land rights and demonstrating that nature is intertwined with power, governance and politics (Braun, 1997). However, the long relationship of Canada's Indigenous peoples with the land since time immemorial provides a source of deep knowledge by which to understand the complexity of our natural environment and its systems as cultural landscapes. A 'database' of ecological knowledge has accumulated from managing the land and natural resources over time. This knowledge, which has been passed down from generation to generation through oral narrative, possesses a high level of meaning and understanding (Pierotti, 2011); it is simultaneously an alternative and complementary expertise to the broader global perspective of Western science (Berkes, 1999; Pierotti, 2011).

Traditional knowledge rises above generalities through local specificity and comprises a world view that sees humans as integral to Nature, rather than separate from it (Berkes, 1999; Pierotti, 2011; Turner et al., 2000). Learning from knowledge accumulated over time through expert observation is fundamental to TEKW (Berkes, 1999; Pierotti, 2011; Schorcht,

2002; Turner et al., 2000); observation leads to a meeting of science and poetry (Angel et al., 2016). Traditional knowledge invokes the spiritual or sacred dimension of ecological system— as well as one of inherent beauty. Ecological wisdom, embraced as "both individual and collective knowledge" (Xiang, 2014: 67), expands by sharing experiences and diverse understandings of relationships to local environments. Wisdom is also gained by reflecting on this knowledge across generations.

Cultural landscapes: poetics of place

TEKW offers a relational world view where every factor is linked and interconnected, with relationships revealed in spoken language and ancestral names; stories, inscribed as literal 'land-marks' that join people to sacred geographic features. Landscape is both a repository of knowledge and a stage upon which people gain experience and learn through intergenerational transfer and continuing practices. The centuries of co-existence with nature have shaped the cultural identity of many Indigenous peoples around the world; cultural landscapes are living landscapes that have been identified as fundamental to these cultural heritages.

Landscape, ever-changing, reflects a continual process of becoming. The experience of landscape is both composite and heterogeneous, requiring both knowledge and memory to appreciate, much less understand, the complexity of place. Key identifiers of the cultural landscape—such as place names, sacred vegetation, preferred vistas or the sound of rushing water—support cultural practices and reinforce our connections to landscape and place. A sense of belonging to a place grows from ideas, memories, cognition and narrative as much as primary senses such as sight, sound and smell. A cultural landscape then is a way of both seeing and being (Fairclough, 2002). It is not just the land, not just the scenery, but what we construct as we shape the world through our senses, our individual and collective memory and narratives, and personal cognition, all communicated between generations.

A large part of what shapes the determination of place and our reading of landscapes is based on commonplace aspects or those of everyday significance. Often significance derives from local, small-scale, non-monumental—perhaps even mundane—things that acquire value by their familiarity and sense of 'fit' with their locality. Their values are intangible and associative. Yet the landscape is also coded with information that conveys previous use, factors integral to understanding a cultural landscape and a people's relationship to it. There may be evidence of bark peeling (an indication of wood harvesting) or of areas managed to control forest fires. Such markers provide visible traces as to how the land has been governed and managed, thus generating an aura unique to the place. Ethnic traces imbued simultaneously with analytical knowledge and emotional spirit, can inform us about culture and the knowledge accumulated over generations of those who have inhabited the land. Landscape and heritage delicately connect and interweave the past with the future, and forge a storied relationship with land (Larsen et al., 2017). Cultural landscapes embody a holistic relationship to land, a relationship that extends beyond the tangible surface of the earth, in a place where cultural, ecological and spiritual values intertwine. Belief systems, concepts of time, movement and world view are all integral components for creating a landscape both cognitive and tangible.

Protecting nature and the cultural landscape of northern Canada

Northern Canadian communities stand on the front line of dramatic environmental change; both cultural identity and actual survival are at stake (Van Loon, 2012). The landscape of the north continues to change, with constant pressures for extreme development at the expense of self-sustaining Indigenous communities. Yet, Indigenous ecological knowledge can provide insight into how we might manage sustainable relations (LaDuke, 1994) within the expansive Canadian Northern territories of Nunavut, Northwest Territories and Yukon.

The northern Canadian landscape has been written—and re-written—over time through treaties, land claim agreements and governance structures imposed and coerced by the federal state. Historically, every act of measurement has been, in fact, an act degrading and disempowering Indigenous culture. The economic and administrative goals of the Canadian government have simplified land tenure and Indigenous community structure, while harnessing nature as an exploitable resource. Land claim settlements have been a formalised process to manoeuvre ownership and transform Indigenous lands, thereby reducing the available land needed to survive culturally and sustain a healthy subsistence-based economy. Maintaining a way of life becomes more difficult to achieve on areas of land of significantly reduced dimensions. Any removal of land is no different than a removal *from* the land and results in a consequent loss of culture.

TEKW: designing new relationships with nature

The North is an exemplary region characterised by dramatic transformation that demands changes in the ways we conceive and design the built environment in relationship to natural factors such as climate and the socio-economic realities affecting the people. Many projects (e.g.

designs for housing or communities) planned in the South are imposed on the North, and fail to respond to the realities of the northern climate, human capacity and challenges of implementation. In all, they ultimately impede sustainable ways of building and living. Often the product of inadequate oversight, the prioritisation of development interests—often dictated by southern Canada and even other parts of the world—over the conservation of nature, erode the cultural landscape and its heritage of place. As a consequence, the 'original' landscape, rare and precious, loses both its economic worth and its value as beauty.

A more responsive and dynamic framework needs to emerge for envisioning Northern communities—one that differs markedly from the fixed visions of Southern planning models—to enable and renew culturally sustainable relationships with the land. Implementing Canadian design and development from a 'North of 60' perspective requires an understanding of a heterogeneous North to guide design and development that respects Northern cultural relationships to land and supports the dreams and life-ways of Northerners. Designs need to better respond to both the tangible and intangible cultural landscapes; designs that ultimately must be informed by multiple voices and knowledges.

I argue that to achieve these goals first necessitates a focus on the preservation and conservation of the North's cultural landscape. Regional strategies should articulate priority zones, identify areas most sensitive or at highest risk, in order to emphasise renewal, mitigation or stewardship. A land-use framework might then determine individual community development plans that reflect local needs and future possibilities. Landscape features, unique ecosystems, rare species, waterways, cultural landmarks, memories, narratives, built form—all need to be comprehensively documented and identified, and mapped by methods other than those laden with colonised boundaries. The next step would be to establish national directives for the protection of Northern cultural landscapes and to prioritise Northern development that does not ignore the integrated presence of the natural and constructed, the physical and spiritual. Such actions would acknowledge the world view of Indigenous peoples and the significance of the cultural landscape as a framework that must inform subsequent design. New development would then better acknowledge the importance of both history *and* future, rather than struggling against or outright rejecting either one or both.

The integrity of the natural environment is critical to the health of communities; this includes collective and individual heritage. These are matters of local, territorial, regional and national concern. The North can offer opportunities for innovative design models rooted in local knowledges and responsibilities for an ecological way of life. Canada is in a position to imagine a future not yet imagined, one that embodies the ecological wisdom of Indigenous peoples.

Designing for new relationships with nature

Landscape design needs to both absorb and enable change; to be flexible and adaptive, without destroying cultural history of the place or its roots. With indigenous expert management and stewardship practices, and from the conscious shaping of the landscape—tied to a regard for traditional knowledge—poetic landscape qualities and authentic sensibilities will evolve.

A shared vision, developed with local expertise, will re-imagine a future with a land characterised by reciprocity and respect, to maintain productive and experientially rich cultural landscapes. Such strategies must be developed from the understanding that although places are culturally and ecologically diverse, they are always in a state of transition. Understanding the dynamic aspects of local ecologies and innovative design models adapted to the landscapes undergoing change, will be required to re-establish landscape integrity, and to restore meaning for that land. If we truly care for a place, we can create dreams and visions for the future that respect histories and cultures, and that yield a renewed, more interconnected, relationship with Nature.

NOTES

1 The *European Landscape Convention* identifies cultural landscapes as natural, rural, urban and peri-urban. They include land and inland water; they are productive, everyday and degraded. Landscape itself is understood as a cultural construct. See Graham Fairclough, "Archaeologists and the *European Landscape Convention*", in Fairclough et al., 2002, p. 27.

REFERENCES

— Angel, J. et al. (2016). The Social Life of Water—Experiential Learning Evaluation of an Intercultural Art–Science Exhibition. Poster presented at Okanagan Research Forum: Building A Resilient Okanagan Landscape, UBC Institute for Biodiversity Resilience and Ecosystems Services (BRAES) and the UBC Institute for Community Engaged Research (ICER), Kelowna, British Columbia, December 2016.

— Berkes, F. (1999). *Sacred Ecology: Traditional Ecological Knowledge and Resource Management*. Philadelphia, PA: Taylor & Francis.

— Braun, B.W. (1997). Buried Epistemologies: The Politics of Nature in (Post) colonial British Columbia. *Annals of the Association of American Geographers*, 87(1), 3–31.

- Fairclough, G. (2002). Archaeologists and the European Landscape Convention. In G. Fairclough, S. Rippon & D. Bull (Eds.), *Europe's Cultural Landscape: Archaeologists and the Management of Change* (pp. 25–37). Brussels, Belgium: Europae Archaeologiae Consilium.
- LaDuke, W. (1994). Traditional Ecological Knowledge and Environmental Futures. *Colorado Journal of International Environmental Law and Policy*, 127, 127–148.
- Larsen, S.C., Johnson, J. & Wildcat, D.R. (2017). *Being Together in Place: Indigenous Coexistence in a More than Human World.* Minneapolis, MN: University of Minnesota Press.
- Patten, D.T. (2016). The Role of Ecological Wisdom in Managing for Sustainable Interdependent Urban and Natural Ecosystems. *Landscape and Urban Planning*, 155, 3–10.
- Pierotti, R. (2011). *Indigenous Knowledge, Ecology, and Evolutionary Biology.* New York: Routledge.
- Schorcht, B. (2002). *The Storied World of Harry Robinson: Emerging Dialogues.* BC Studies, 135, 145–162.
- Turner, N., Boelscher Ignace, M. & Ignace, R. (2000). Ecological Knowledge and Wisdom of Aboriginal Peoples in British Columbia. *Ecological Applications*, 10(5), 1275–1287.
- Van Loon, R. (2012). Challenges and Expectations in the Canadian North. In F. Abelle et al. (Eds.), *Northern Exposure: Peoples, Powers and Prospects in Canada's North* (pp. 529–560). Montreal, Quebec: The Institute for Research on Public Policy.
- Xiang, W.-N. (2014). Editorial: Doing Real and Permanent Good in Landscape and Urban Planning: Ecological Wisdom for Urban Sustainability. *Landscape and Urban Planning*, 121, 65–69.

(1 May 2018) Today the pear trees are in blossom and HUGE bumblebees are staggering around about 20cm from the ground. Drunk. Or cold perhaps, a little stiff…

Sarah Pennington, Eleanor Margolies

Antarctica SE3: A Conversation on Designing with Care

The year 2016 was historically unprecedented, with a record global temperature and exceptionally low sea ice. According to a report from the World Meteorological Organization…

We talked about the difficulty of reading a text that begins this way.

In the summer of 2016, in the South Pole Observatory in Antarctica, levels of carbon dioxide hit 400 parts per million for the first time in 4 million years (Kahn, 2016), the last and most remote monitoring station on Earth to reach this figure.

In London, where we live, it was the hottest September since 1911.

Could we take these climate facts and 'thicken them' with care?

A heatwave can be described as a more-than-human entanglement: an assemblage of temperature, humidity and pressure interacting with geology and the pattern of land use, as well as local practices of agriculture and architecture. Heatwaves are a manifestation of the extreme weather events now increasing in frequency and severity across the globe, leading to crop failures, forced migration, deaths and the destruction of ecosystems. The effects of extreme weather are unevenly distributed—one indication of the complexity of climate change.

From our situations—as a design researcher and a performance researcher—we wondered how to pay attention to these more-than-human things of ecological concern: gases and liquids, melting and burning? We wondered how to 'live with' the entanglements of climate change: the links between Antarctica and London. We were looking for a way to talk about a particular "matter of concern" (Latour, 2004) which often seems muted or silenced in conversation (Marshall, 2014).

This part of southeast London had been home to three early polar explorers. By linking their homes, we created a walk for considering this entanglement, an 'Antarctic expedition' beginning in postcode SE3. Invitations were sent out widely and ten people with a diverse range of knowledges and relations to the neighbourhood joined the walk: designers, artists, writers, a political activist, a community gardener, a transport infrastructure manager, an IT manager. They were each invited to bring along a 'material of climate change' as a starting point for conversation. Yellow woollen hats and 'Antarctica SE3' badges gave this group of strangers a temporary collective identity, a licence to be playful in their exploration of the neighbourhood and the ecological concerns.

The event was conceived of as a series of conversations, non-didactic, intimate and open-ended. Participants walked and talked in pairs, swapping over at key landmarks—a car showroom, a petrol station, an ice house.

Movements 1–5 (of 9) Sunday 18 September 2016

#1: Walk over the heath. Stop by the plaque.

Blackheath, SE3. The heatwave had scorched the expanse of grass brown, revealing bare ground and dusty soil. A flat, open landscape; around the perimeter of the common are large houses, trees and vehicles moving on hot roads; in the distance, the City. On the front wall of a nearby house, a blue plaque: *Sir James Clarke Ross, Polar Explorer lived here, 1800–1862.*

#2: Participants share materials of climate change.

Someone showed an iPhone; another talked about the extraction of rare-earth minerals used in the production of technological products; someone else brought a T-shirt from the Florida Keys, where 90 per cent of the land mass is just five feet or less above sea level; an electricity pre-payment key that 'makes the cost of energy tangible'; a bottle of English wine had been brought to show that 'perhaps global warming is not all bad'.

#3: In partners, walk and talk for 1.2 miles south to Manor House Gardens.

#4: The Ice House. Meet the caretakers.

Down into the cold of a stone-built ice house, with two store chambers and a domed ice-well, used to preserve food before refrigeration. Built in 1773, it held ice from local ponds and later, ice shipped from Scandinavia or from the Hudson River in North America. Even without any ice, the temperature underground was significantly colder than the heat outside; a sensory evocation of subzero polar temperatures.

#5: Walk south, stopping by the shipping containers on the heights of Mountsfield Park to eat sledging biscuits. Rather like unsweetened shortbread, these are made of flour and butter and are used to replace bread on Antarctic journeys.

Antarctica SE3 participants. Image credit: Gemma Lord

Reflecting on the project, we found that it opened up some questions around how to make 'with care'.

We started by designing a way to consider a matter of concern.

We wanted to understand why concern required something more, needed to be 'thickened' somehow.

We found we wanted to thicken concern with care.

We wanted to understand what it would mean to 'care with' rather than 'about' or 'for' the non-human.

Maria Puig de la Bellacasa, in her recent engagement with Bruno Latour, has been a guide here. As she describes it, care is something more complex than a moral cliché or a smoothing out of differences. For her, care is a practical, affective and ethical act, that involves a speculative commitment to 'assembling neglected things' (2011). By encouraging our attention to the neglected, she broadens out the scope of caring from human and non-human care, to the more-than-human (2017). In our case, a heatwave.

Puig de la Bellacasa (2012) also draws our attention to 'thinking with care' as a sensitising ethos. As a relational way of thinking, care also affects the way we produce knowledge. It is an ambiguous term, since care — or the lack of it — can be expressed in different ways, depending on who is describing it; and where caring for one thing may mean *not* caring for another, highlighting the contested nature of care (2017: 9). Thinking with care therefore invites us to add layers of meaning and 'hold up the multiple ends of sometimes divergent positions' (ibid.: 72).

During Antarctica SE3, we didn't summarise or resolve contradictory viewpoints. This would have been a reductive act at odds with the liveliness of care. Our decision not to resolve things didn't mean setting up a debate; instead, we had an ambition to promote conversation.

In some fields, the fierceness of a discussion is often taken to indicate its quality. Debate is promoted as a positive value in, for example, the 'battle of ideas' in politics or philosophy; and in 'design for debate', design outcomes are placed in public arenas to generate debate on issues of concern. Conversely, in participatory design projects, consensus is often an end-point to be achieved (Keshavarz & Mazé, 2013). However, Nel Janssens (2017: 152) suggests that we reconsider the conversation as an alternative, to shift the emphasis from "friction" to "bringing together", from "exchanging arguments" to "sharing experiences".

A number of features of our event enabled this — including the series of one-to-one exchanges between participants, not structured around climate-expert input but rather on responses to certain sites, objects and interactions.

We also wanted to design a format that offered a way of 'living with' climate concerns. We began by discussing our experience of participating in events that had been designed to direct attention to the environment, or to create a deliberate shift in the relationship between human and non-human. We were both drawn to the notion of a walk as a designed event in which the design work might be unostentatious, a reconfiguration of a socio-material setting rather than the creation of an extraordinary new design object. Designing such a walk would involve mapping, research and testing while allowing space for contingency and chance. This approach to design potentially models an attitude to nature that is appropriate for the present moment: engaged and responsive, recognising the intertwining of what Guattari called the "three ecologies"—environment, human subjectivity and social relations (Guattari, 2000).

Walking has been described as a "social way of thinking and knowing" (Ingold, 2010). This sociability is engendered through a "common movement" when walking side by side with others (Lee & Ingold, 2006). Walking has also been described as a speculative mode of engagement (Gabrys, 2017) as there is always an element of unpredictability about what's around the corner. A walk as a format therefore seemed germane to our situation. It provided a relational format for collective engagement, grounded in the embodied experience of a heatwave in London, and could also promote speculation on the complexity of climate change through an imagined connection to a polar expedition.

We placed these experienced and imagined locations—the London Borough of Lewisham, and Antarctica—in relation to each other, and set about to design an event that would 'remix' our relationship to a heatwave through interaction with the mundane infrastructure of paths, commons, cultural sites. We paid attention to signs of the carbon economy in our neighbourhood that we normally overlook, such as petrol stations and shopping centre car parks. At some locations, historical layers are officially recognised by blue heritage plaques to explorers, and information boards giving accounts of local history while at other times we found ourselves in 'edgelands', walking along hidden rivers and footpaths unknown to Google Earth.

> *To say 'I started a conversation' sounds quite flippant. But it isn't. Making a space to start a conversation is really important.*
> (Participant feedback)

At the end of the walk, gathering for a discussion in an upstairs room, the intimacy of the conversation in pairs dissolved. We found ourselves wanting to challenge statements, and change minds—falling back into more usual rhetorical patterns that had been set aside during the walk. Nevertheless, as participants shared thoughts and comments from their walking partners, views that were not necessarily ones they held, the gathering continued to 'hold' a proliferation of viewpoints rather than moving to a shared position.

Participants individually recorded their thoughts on the experience, as well as continuing discussion in an email thread after the event. The themes of care, of conversation and walking, emerged in their reflections.

We had looked for neglected things in order to thicken climate facts; to make 'with care'. Participants commented on how they had discovered overlooked issues in familiar settings, and how a shared attention to the environment they were moving through allowed for productively tangential approaches to the declared topic of climate change.

One participant noted that the walk had created "a different space where conversations happen". The early moment in which participants showed their 'materials of climate change' exposed diverse opinions on the issue, as well as enabling participants to choose their initial conversation partners. The conversations were open-ended, sometimes looping back in reference to a previous discussion. Expertise was temporarily set aside to allow for the sharing of experiences and emotions. As one participant put it, rather than learning from one person's expertise, there was an interaction of knowledges, "like a collection". With participants bringing divergent positions on climate change, the non-directive format allowed for what Jacques Rancière called a mutual "lodging of worlds" (quoted in Keshavarz & Mazé, 2013).

Part of our objective for Antarctica SE3 was to make a work that was a participatory engagement with the current ecological crisis. Walking together on a hot September day located these multiple, simultaneous conversations in a remix of experienced and imagined locations, of past and present, of the human and the more-than-human. Through the walk, it became possible to pay attention to the non-human: visually, as we took in the parched heath; acoustically, noticing the difference between main roads and tree-shaded footpaths; physiologically, as bodies responded to the heat. We gave ourselves a chance to hear the more-than-human as one of the many unresolved voices participating in the walk. Stepping into the project of polar exploration brought into focus the legacies of the colonial exploitation of resources; we imagined Scandinavian ice in Lewisham a hundred years ago, and petrol burning in London melting polar ice-caps: melting and burning.

It will be clear that this walk, both due to its location in London and in the self-selection of those who chose to join a speculative walk on climate change, is far from representing a full spectrum of the voices that need to be heard on climate change. But we suggest that it might be one way to make a space to start a conversation. Making 'with care' implies situated and relational forms of engagement that foster a generative speculation on neglected things, including the more-than-human.

REFERENCES

- Gabrys, J. (2017). Air Walk: Monitoring Pollution and Experimenting with Speculative Forms of Participation. In C. Bates & A. Rhys-Taylor (Eds.), *Walking Through Social Research* (pp. 145–162). London: Routledge.
- Guattari, F. (2000). *The Three Ecologies* trans by Ian Pindar & Paul Sutton. London and New Brunswick, NJ: Athlone Press.
- Ingold, T. (2010). Footprints Through the Weather-world: Walking, Breathing, Knowing. *Journal of the Royal Anthropological Institute, 16*, S121.
- Janssens, N. (2017). Collective Sense-making for Change: About Conversations and Instructs. In M. Schalk, T. Kristiansson & R. Mazé (Eds.), *Feminist Futures of Spatial Practice: Materialisms, Activisms, Dialogues, Pedagogies, Projections* (pp. 151–158). Baunach, Germany: Spurbuchverlag.
- Kahn, B. (2016). Antarctic C02 Hit 400 PPM for First Time in 4 Million Years. *Scientific American* [online]. Retrieved From https://www.scientificamerican.com/article/antarctic-co2-hit-400-ppm-for-first-time-in-4-million-years/
- Keshavarz, M. & Mazé, R. (2013). Design and Dissensus: Framing and Staging Participation in Design Research. *Design Philosophy Papers*, issue 1.
- Latour, B. (2004). Why Has Critique Run Out of Steam: From Matters of Fact to Matters of Concern. *Critical Inquiry*, 30.
- Lee, J., & Ingold, T. (2006). Fieldwork on Foot: Perceiving, Routing, Socialising. In S. Coleman & P. Collins (Eds.), *Locating the Field Space, Place and Context in Anthropology*. ASA monographs 42. Oxford: Berg.
- Marshall, G. (2014). Don't Even Think About It: Why Our Brains are Wired to Ignore Climate Change. New York: Bloomsbury.
- Puig de la Bellacasa, M. (2011). Matters of Care in Technoscience: Assembling Neglected Things. *Social Studies of Science, 41*(1), 85–106.
- Puig de la Bellacasa, M. (2012). 'Nothing Comes Without its World': Thinking With Care. *Sociological Review, 60*(2), 197–216.
- Puig de la Bellacasa, M. (2017). *Matters of Care: Speculative Ethics in More Than Human Worlds*. Posthumanities 41. Minneapolis, MN: University of Minnesota Press.

Kate Fletcher

Folding In

Since early spring I've been eating every meal outside. This is a habit I've been cultivating for the last five years since I spent a fortnight on board a sailing boat in the Western Isles of Scotland. On the boat I couldn't bear to miss a thing and so I decamped to deck, the outdoors becoming the dining room of choice. Truth be told, I also couldn't see a thing below. In the space of two days, I went from being blissfully sharp in my focus of the world both near and far, to being unable to see myself in the mirror. I couldn't thread a needle, read a word. I was suddenly longsighted. Caused perhaps by gazing endlessly at the distant ocean (later I found out that this happened to many others before me). The sky was luminous midsummer. Seabird colonies dripped from island cliffs. The ocean writhed with whirlpools and showed dolphins, phosphorescent creatures, fish. When I got over the awkwardness, I borrowed a spare pair of the skipper's ready specs. But the sea's legacy is, for me, a permanent physical reminder of the need to look long. To seek out other perspectives, underlying patterns, gathering relationships and subtle changes in surroundings; and this in order also to better understand the conditions close by. I like to think that I might have always done this, but longsightedness has leant this task an uncommon edge, for it is now the clearest vision I have. And eating outside helps.

My lunch today was a sandwich and I ate it whilst sitting on a folding deckchair in the narrow corridor of wildflower meadow at the back of our house. The meadow is easily the best gardening I have ever done. On poor soil I sowed a mixture of native flower and grass seed in handfuls from a deep fold in my skirt. As I did it, I caught myself with a laugh, a moment of consciousness of taking part in an age-old task: shaping my arms and broadcasting seed as humans have for millennia. It's the same feeling I have when dragging bundles of firewood, firing an arrow. A mix of instinct and recognition. I almost ran upstairs for a scarf to tie back my hair.

As I ate I watched the bees. There are so many this year, especially bumbles, that watching them leaves me feeling slightly frantic. Early in the spring, the meadow was mainly flowered with ribwort plantain. Plantain throws up tan-coloured slender tubes, rising to about knee height, each with white petals in a halo. It grows in cracks in the pavement, on waste ground; it's not exactly a plant to write home about. But for bees, it's the cat's pyjamas. The bees foraged the plantain obsessively, working the stalks round and round, their chaps bright with pollen. That was in May and at lunchtime today, early July, lesser knapweed was the total draw. Knapweed stands in big purple clumps, at eye height as I sit in the deckchair. Its flowers have a shape a bit like an approachable thistle. In this meadow I reckon there must be one bee for every three knapweed flower heads. It is a mass communal feast to which I've only brought a sandwich. As I see it, knapweed is nice, but this intoxicating? Really? My nose can detect no scent. Its flower heads are compact, perhaps this makes a firm base on which land, a stable platform? Its purple colour is a little short on blue tones to be a true purple, its centre leans towards pink. Is this the lure? Whatever it is I realise how scant my understanding is when it comes to the tastes of bees. It's almost like there's a rule of inverse proportionality at play. The less I notice a flower, the more the bee is smitten with it. This rings with counter-intuitiveness, of the need to not blindly trust my own tastes, but to follow the opposite feeling. It throws everything up in the air. But perhaps that is what is required. Who knows how things will land as we build a counter narrative to the grand narrative of our times; one based in a renewed connection between humans, places and the beings that dwell in these places.

I kept on with my lunch and watched a honey bee. It rifled through the stamens on a knapweed flower. And then a tiny distance above it, I saw a hoverfly. The hoverfly measured about half the bee's size and wore wasp hazard stripes. As the bee moved to a new flower, the hoverfly followed. There was a tiny lag, like a reluctant shadow and a momentary upwards bounce in the hoverfly's flight, before it resumed its steady holding a thumbnail's width above the bee. And this was how they carried on. Bee to flower. Hover to bee. I wondered what was going on. My first thought was that they were they friends. And then, was the hover a pet? Or maybe a parasol, a portable fan? Then I realised I might have this the upside down. Was the hoverfly in charge here, an overseer? And the bee its worker? The duo moved on to a new patch of knapweed. And then the two became four. Two more hoverflies joined the holding pattern, stacked vertically. One bee, three hovers. It looked like a diagram of the layers of middle management. In the time it took to eat a sandwich, again I realised how little I know. It is as Annie Dillard says, a thing such as this, 'will never make sense in our language but only in its own, and that we need to start all over again on a new continent, learning the strange syllables one by one' (1974: 108). I have to learn different sounds, fresh

sights, a new lingua franca. Perhaps only then might I speak the language of nature's agency.

As I went back into the house I remembered a quote I once read, 'Nature is not only more complex than we think. It is more complex than we can think' (Egler in Barber, 2014: 88). Nature is beyond the comprehension of the thinking brain. But it is not beyond us. It is us, it is the wilderness parts of our imaginations, the untamed parts, the loose parts, the intuitive and sentient and inventive parts. It is all of the same parts where design thinking and practice have their roots. But to work from this untamed place within us takes practice, perhaps what Gary Snyder calls the 'practice of the wild': 'A deliberate and conscious effort to be more finely tuned to ourselves and to the way the existing world is' (1990: viii). To notice, to revel in what we find and then to act. Only by bringing things into being with our hands, our tools, our voices and words, do we continually participate in the on-going renewal of life in the world. For me this act of participation feels like blending and combining; like the folding in of flour to creamed butter, sugar and eggs in the making of a cake. Folding in is how it feels to me.

On 24th April 2013 the Rana Plaza clothing factory in Bangladesh collapsed and more than 1000 garment workers lost their lives and 2500 were injured. I, like many others, felt cleaved apart by the tragedy and what the factory collapse said about the skewed priorities of the fashion system, about how in unplanned combination they have contributed to an unconscionable way of producing and consuming clothes. The system, the industry in which I work to design alternative systems and practices, has no care for those it is dependent on. I was both angry and ashamed.

On the first-year anniversary of the disaster I watched an interactive film about the tragedy and spent time on social media campaigning for change from a tiny cottage in which my family and I were then living.

The cottage was part of a farm on a bluff at the edge of Macclesfield Forest in the north of England. Outside, pheasants barked, light gathered in pools and the wide and deep landscape revealed it all: change and permanence, the objective reality of our planet.

The cottage was about a mile from the nearest road and the lane that linked us was steep and full of bends, like a fairground ride. At the lane's edge was coarse grass and broken drystone walls. Brown hares lived there. They ran along the lane. They legged it in the late afternoon, at dusk and at night in the tunnel of light thrown out by our car's main beam. Their movement was pleasure and ease, the hinging and folding of limbs, feet, independent ears swivelling, listening. One evening as I drove home, I watched one streak ahead of our car and then for no apparent reason, she stopped dead. She sniffed the air and turned. We looked at each other, face to face, I did not blink. Slowly I eased the car past. I watched her in my rear view mirror return back the way she came. She lived as she should, as I should, true to her experience; choosing her route with a keen and fierce will.

After watching the documentary, my head still scattered by it all, and when my two sons were home from school, I went with them down to the tree-stump den that they had been building in the forest. We followed a path made by deer, through larch and sycamore. The den itself was a muddle of wind-blown sticks and we swarmed around it, building, messing about. We dragged branches, got our hair tangled in twigs, and passed time lifting things, shoving things, idly between the trees. In jackets globbed with pine resin, we added more branches. A flock of tiny birds moved in the tops of the trees. A bank of rain edged closer.

It is as Hasebe-Ludt et al. say, "The stories of our lives, if only our own stories, would not be worth telling. These stories become an ethos for our times, as they expose interdependence and interrelatedness with all beings" (2009: 129). Things exist only insofar as they can be related to other things: trees, the rain, a broken fashion system; they intermingle human interests with all others'. The deadliest structural failure accident in human history is an issue that is not supply chain alone in its reach and shape. It is not just industrial, commercial. Nor are the questions at stake only those about factory workers. They have deeper roots that tangle with western culture's self-enclosed outlook; with its tendency to instrumentalise both ecosystems and workers as servants or resources, as a means to the fashion system's ends; and with the view that nature is the inessential, unconsidered background to technological society. Sixty years ago, Aldo Leopold put it this way: "We fancy that industry supports us, forgetting what supports industry" (1959: 178). The human tragedy of Rana Plaza—perhaps all the more relatable because people are involved—is a reminder that, in order to protect workers and everyone else, we need to see the whole, to ecologise the fashion system. Our task becomes one of embedding our work in the systems it is dependent on, of growing our understanding of it as part of an interrelated earth so as to nurture an expanded community of relations on which all lives depend. I think this is what Val Plumwood meant when she advocated for, "a much larger, less humanized community, with an ethic of respect and attention needing no stopping point" (2013: 443); a community where humans are not the focal point nor the limit of care, respect and attention. After all, as Plumwood states, "human-centredness is not in the interests of either humans or non-humans" (ibid.: 443). We are all in this together.

The den, the hare, the flowers, the bees, are snickets into a world that exceeds us, that is greater than our potential and our knowledge. This world, our world, draws us outwards and onwards. Here the currency of choice is the capacity to care, to give attention, to feel the attention returned. Perhaps then things like the siren song of self-interest, the part of fashion that gets all the play and all the headlines and is marketable and profitable, gets reframed in our minds in the process. Fashion is not just an expression of human-centred individualism. It is also about solidarity with all our ecological relations. Most fashion experiences neglect getting support from others. But in folding in the agency and independent value of our relations, fashion experiences become as much about reciprocity as self-interest. This is the long view; it can open up a wild, untrammelled mutualistic practice of design, a counter narrative of connection and care, where fashion takes us towards the earth.

REFERENCES

— Dillard, A. (1974[2014]). *Pilgrim at Tinker Creek*. Norwich, UK: Canterbury Press.

— Egler, F. in Barber, D. (2014). *The Third Plate: Field Notes on the Future of Food*. New York: The Penguin Press.

— Hasebe-Ludt, E., Chambers, C.M. & Leggo, C. (2009). *Life Writing and Literary Métissage as an Ethos for Our Times*. New York: Peter Lang Publishing.

— Leopold, A. (1959). *A Sand County Almanac*. London: Oxford University Press.

— Plumwood, V. (2013). Nature in the Active Voice. In G. Harvey (Ed.), *The Handbook of Contemporary Animism* (pp. 441–453). Durham, UK: Acumen Publishing.

— Snyder, G. (1990). *The Practice of the Wild*. Berkeley, CA: Counterpoint.

Conclusion

Black backed gulls in the cold winter rain, a tall White Alder, Siberian jays, a lanternfish, harakeke of New Zealand, urban moth, a boy in an itchy polar bear costume, pollinating bees, a touring calathea with electrodes attached, black volcanic sand, a girl enchanted by chaos, fungi spreading in the kitchen, a ground squirrel.

These are some of the figures from the book that stay with us editors. These guides, muses, inspirators and trouble makers are spirits that have been presenced through stories, imagery, poems. They help make the emergent field of design and nature rich, nuanced, complex, dirty, alive. The figures simultaneously act centrifugally to bring our questions, attention and energy to a shared core ethos of design and nature, and ground them in a core within ourselves. They also act centripetally throwing us outwards into a myriad of new possibilities, landscapes and sites, temporalities, and ways of relating. It is exciting and humbling. We are grateful for the gifts these figures and the contributing chapters bring to design and nature.

For us, Kate, Louise and Mathilda, it has been meaningful to also offer four posture and modes—lying, sitting, standing and walking—as ways of being with design and nature. We hope they can inspire a multitude of design and nature practices. Deborah Bird Rose has spoken of life's capacity to offer intergenerational gifts, and how this is threatened because of disastrous processes of our time. We consider the contributions of the authors as a web of gifts to practice, research and education. The gifts are not all new. Some reinvoke past and ancestral practices and knowledge; some are about the unmaking of current practices.

> *The kiss of life is an ancestral blessing, alive, brilliant, and pulsating in the world around us and within us… We are called to acknowledge that in the midst of all we cannot choose, we also make choices. And we are called into recognition: of the shimmer of life's pulses and the great patterns within which the power of life expresses itself. We are therefore called in gratitude for the fact that in the midst of terrible destruction, life find ways to flourish, and that the shimmer of life dies, does indeed include us.* (Bird Rose, 2017)

Making a book takes those involved on a journey. Since we set out on this book adventure in 2015, the discourse on more than human worlds has expanded. Terms, such as the Anthropocene and speciesism, have become part of a design vocabulary and texts such as Donna Haraway's (2016) *Staying with the Trouble: Making Kin in the Chthulucene*, have begun to infuse design. Indeed, one of the first events of 2019 attracting design researchers was a Multispecies Storytelling conference at Linnaeus University, Sweden.

We are grateful for these emerging discourses and networks, and hopeful that they are coupled with new practices. We also feel a deep sense of urgency. Explorations of design and nature must not stay as beautiful, disturbing or interesting ideas, they need to permeate design of things, processes, habits and practices, infrastructures and legal frameworks, cultures, norms, languages and paradigms.

Since so much of the dominant understanding of design is intertwined with the modernist project, growth logic and capitalism, extraction of natural resources, a Western hegemony, and patriarchy, this means that a genuine exploration into relations between design and nature almost inevitably involves losing our design footing. Who will we be, what will we do, what will our contributions be in a new paradigm of design and nature relationships?

In Louise St. Pierre's historical contextualisation of design's relations with nature, keywords are mastery, control, technological determinism and optimism, capitalism as well as romanticism. In hindsight, human centredness and illusions of human control have persistently shaped design's many attempts to relate to nature. This leads us to ask: what are key traits of design and nature relations coming out of this book; how will these narratives be seen in a future? Is it even possible that we humans can ever imagine ourselves away from human centricism, away from illusions of being in charge? Louise asks designers to change the underlying terms of our historical relationship to nature.

We are grateful and humbled that on this journey of making the book we have encountered so many design scholars and practitioners willing to let go of stale alliances and instead find new partnerships in doing, ways of knowing, understanding our world. The new epistemological positions that emerge notably embrace queer and feminist perspectives that, for example, challenge linearity and conformity, and bring practices of repair, tending, maintenance, care into the remit of design, as well as 'simply' holding and attending. They challenge the solutions focus of dominant design practice and, as significantly, the viability of design stars and heroes. Responsibility and action for healthy design and nature relations must be distributed, as no one organisation or individual can hold the problems or possible pathways ahead.

The contributing chapters bring many gifts, to design research, education and practice. Some that we can discern are:

A deep connection with plant, animal, place, land, *here*. The 'super-situatedness' of several of the contributions is certainly a turn from standardisation and perfection. It also creates a learning loop where inspiration, exploration, material, processing and use feed each other. They are open to emergent and meaningful process and outcome. Here, a definition of creativity is also challenged, as authors critique a model of design that forces a desired shape onto nature. An assertive programme for design, broken down in clear steps, is therefore replaced with guiding qualities or principles, such as sensing, attuning, relating and reciprocity.

Several contributions highlight the friction of design and nature relations, both historically and moving forward. They point out that there is risk, bite, grit and often an awkwardness to these relationships, and that exploring design and nature also requires examining design and nature relations within ourselves. In this way, the book traces pathways for design and nature void of romanticism, nostalgia and the twee. Taking design's relationship with nature seriously, in terms of leaving behind power imbalances, requires entering it as a mature individual, and considering nature a mature partner in dialogue and dance. This space therefore avoids cultivated naivety and fetishism, but does comprise awe and wonder. An important point of exploration is how we can navigate intimacy and our own vulnerability without reverting to clichés of being close to nature that end up detaching us. How can we lean in, touch and be touched and take responsibility?

What it means to 'know' in design and nature is also about recognising and accepting the limits of our understanding and the potential synergies of many ways of knowing and many knowledge holders. As we stay with the troubles (Haraway, 2016) that our guides gift us, we also start challenging the terms knowing and understanding, central to the academic institution and a logocentric heritage. These terms imply a binary quality—to know/understand or not to know/understand—which does not reflect dynamic relations between design and nature. We propose *learning* and *agency with care* as more helpful pursuits. In the world-making that practising new relations between design and nature implies, play, imagination, speculation, fabulation occupy a central role.

Participative design is now celebrating several decades and the gift of an awareness that a broad range of humans have a stake in design and therefore should participate in design. Fundamentally accepting a paradigm of participation also means accepting a relinquishing of control so as to give space for voices of more stakeholders. This should lead to more equal distribution of ownership, commitment, responsibility, agency. 'Less' control also means more surprise—as regards the focus of given projects and as regards how design itself is interpreted. With a turn to participation of and partnership with multiple species, the challenges and gifts of participation should be multiplied. They require us to consider at much greater depth what it means to be in the world, the limitations of human language, methods and processes, intelligence, ability to grow, live, adapt, die. It also requires that we look at ethical guidelines from other vantagepoints. This is neither a linear transition, nor a turn back. It is instead an interweaving that allows for many strands to be fuzzy to conventional human consciousness, yet still regarded as integral and of dignity.

There are some examples of pedagogies for design education in the book. They point to educational practices situated in the place, land, soil, tree and, of course, season. They give time for reflection in this space, and for transformation and transgression in this space. We imagine the design studio moving into nature, urban, rural and in between; and complementing design lecturers and workshop staff with members of other species. We imagine that design briefs and areas of investigation may not be predefined, but set by students in relation with nature. This will entail a deep attunement and listening to what is being asked by nature. We envisage that such briefs do not concern what human can design for nature, but what may unfold between. We imagine temporalities becoming more central in design education, both to ask what are our intergenerational and inter species learning gifts and responsibilities; and how can we learn with different species' temporalities. We imagine design students learning more about natural systems, and working as a matter of course in teams with a range of disciplines, including biologists. We imagine education drawing on the four postures.

Writers in this book are open to ways of knowing in design research that support the intuitive holistic character of design, countering the scientism that has crept into mainstream design research. We imagine knowledge making derived from what cannot be seen or quantified in the modernist sense will gain credibility and enter a broader discourse. Deeply local, situated research and accessible research is prioritised as the urgency of the environmental predicament becomes lived experience for many. We imagine multiple species collaboration and participation form the basis for research; for funding bodies, research processes, reaching publics. Several researchers have worked with technological interfaces to access knowledge about nature. This locates technology as a means, rather than an end. Digital technologies may have a relationship with nature that is about gaining intimacy, if we retain focus on the power of nature, and remember to respect nature's communications on her own terms.

Maybe future readers of this book will sense a deep fear of ecosystem collapse; probably articulated perspectives of decoloniality and feminism, as well as positions challenging speciesism. We feel there is also a yearning for connection running through these pages and the courage to open up, let go, see what happens, accompanied with a trust that it will be worth it. We hope the book will be helpful in releasing a space of design and nature with many possibilities and opportunities, involving reciprocity, collaboration, vulnerability and as much humility as we can muster. This will be done, at least in part, through love. Through standing in love with nature. A love that transcends the discourse of biophilia, as love for all species and for their inherent value. Perhaps when we speak about such things, it is especially clear that this is an emerging field. Let's enjoy the fumbling qualities, and be grateful for the imperfections.

REFERENCES

- Bird Rose, D. (2017). Shimmer. In A. Tsing, H. Swanson, E. Gan & N. Bubandt (Eds.), *Arts of Living on a Damaged Planet*. Minneapolis, MN: University of Minnesota Press.
- Haraway, D. (2016). *Staying with the Trouble. Making Kin in the Chthulucene*. Durham, NC: Duke University Press.

(10 May 2018) Today is the day of maximal apple blossom in our garden. It is intense and fluffy white. A scent of sherbet.

Contributors

Eugenia Bertulis trained in Industrial Design during the wild years of the dot-com bubble in the Pacific Northwest. She is surrounded by people designing brilliant things for kittens and yeasts and trees. Her research interests include interaction design and affect, and neo-materialist approaches to ethics in design. She seems to know every philosopher yet no one ever sees her reading.

Zach Camozzi's interdisciplinary practice includes playing in the mud (design research), making forts (industrial design) and taking others canoeing (outdoor education). At Emily Carr University of Art + Design he focuses on producing connections to natural environments, beings and phenomena by running as many classes in outdoor settings as he can.

Cameron Cartiere is an artist, writer and researcher focused on public art, urban renewal and environmental issues. She specialises in community-based collaborative projects. Dr. Cartiere is an Associate Professor at Emily Carr University of Art + Design. She is also the coeditor-in-chief of the peer review journal, *Public Art Dialogue*.

Marcus Dénommé grew up on Algonquin Temiskaming first nations territory in so-called Ontario, Canada. Nowadays, he thoroughly enjoys climbing trees among many others around the Coast Salish territories. He is a practicing printmaker and the editor of the publication Veins & Arterys.

Katarina Dimitrijevic is an interior architect, lecturer and design activist. She is currently affiliated with Goldsmiths UoL, Design Department as an MPhil/PhD researcher, funded by the Design Star CDT, AHRC, UK. Katarina leads KraalD, a social waste praxis, committed to promoting design as a vehicle for social and material transformation.

Kate Fletcher is a fashion sustainability pioneer and design activist. She is a Research Professor at University of the Arts London, UK, based at the Centre for Sustainable Fashion. Kate is the author and/or editor of seven books translated into as many languages.

Raune Frankjaer holds an MA in Spatial Interaction Design and is currently a doctoral candidate in Digital Design at Aarhus University, Denmark. Her research on sympoietic technology centres on non-anthropocentric applications of information and communication technologies, with a particular focus on interactive digital artefacts and haptic body-worn interfaces.

Nancy Holmes is a writer who has published five books of poetry and who collaborates with colleagues and artists on community-based eco art projects, most recently with Cameron Cartiere on Border Free Bees. Nancy is an Associate Professor in the Department of Creative Studies at The University of British Columbia Okanagan.

Li Jönsson is Associate Senior Lecturer in Design Theory & Practice at Malmö University. Her designerly research engagement concerns questions around the more-than-human, collaborative futures, and 'other' aesthetics – and aims to open up for alternative ways of understanding, intervening in, and expanding contemporary issues through design experiments.

Rangi Te Kanawa comes from a family of traditional Maori weavers, this background led to a position in Textile Conservation. In 1990 Rangi completed her Bachelor of Science in Conservation Cultural Materials at Canberra University, Australia and interned at the Textile Conservation Centre, Hampton Court and British Museum (1992).

Faith Kane is a design researcher and educator working in the area of textiles and materials. Her research interests include design for sustainability, collaboration in the design/science space and the role and value of craft knowledge. She is an editor of the *Journal of Textile Design Research and Practice*.

Angela Kilford Te Whanau A Kai, Ngāti Porou, Ngāti Kahungunu is an artist and designer living in Wellington, NZ. Kilford's inspiration comes from Māori concepts and knowledge. These ideas are expressed through writing, performance, collaborative making, large scale public installations and walking.

Frederik Larsen is the co-founder of the sustainability consultancy In futurum. He holds a PhD from the Copenhagen Business School (DK) and has done research on sustainable business practices, reuse, second-hand markets and design. He advises companies and organisations and publishes on design, reuse and sustainability.

Tau Ulv Lenskjold is Associate Professor at the University of Southern Denmark. His work investigates design collaborations with disadvantaged groups and collective engagements around ecological issues. Drawing on participatory and experimental practices from co-design, speculative prototyping and activism, the research explores design as an inquiring approach towards social and environmental sustainability.

Anette Lundebye's work explores the boundaries between values, perceptions, practices and social trends associated with material culture alongside a strong interest in creativity and sustainable futures. She is a senior lecturer, a metadesign researcher and a visual artist.

Maarit Mäkelä works as an Associate Professor of Design in the Aalto University, Finland. Her creative practice, that is contemporary ceramics, serves as a basis for her practice-led research. She has studied creative processes and used especially documentation and reflection as a method for gaining insights and understanding from the process.

Eleanor Margolies writes about performance, cities and the senses. Publications include *Props: Readings in Theatre Practice* (Palgrave Macmillan, 2016) and the edited collections *Landing Stages* (2015) and *Theatre Materials* (2009).

Sarah Pennington is an AHRC PhD candidate at Goldsmiths, University of London, and a lecturer in design at Goldsmiths and The Royal College of Art. Her current research seeks to align a feminist ethos of care in relation to speculative design research practices.

Katherine Pogson has worked as a designer maker and educator since 2000. Her current practice-based doctoral research (University of the Arts, London) originates the concept of the 'Companion Object' to interrogate the potential for handcrafted artefacts to open a dialogue about our relationships with nonhuman nature.

Timo Rissanen is Assistant Professor of Fashion Design and Sustainability at Parsons School of Design. He is a Finnish, Australian trained academic, artist and designer. He completed a practice-based PhD on zero waste fashion design at the University of Technology Sydney in 2013.

Tanya Maree Ruka Te Miringa Te Rorarangi of Ngati Pakau, Ngapuhi tribal descent is an artist and designer currently living in Auckland, New Zealand.

Erik Sandelin is a PhD candidate in Art, Technology and Design at KTH Royal Institute of Technology and Konstfack University of Arts, Crafts and Design in Stockholm. Before returning to academia Erik co-founded and ran interaction design and innovation studio Unsworn Industries for almost ten years.

Laura Sansone is a textile designer and activist. She founded Textile Lab, a design company that promotes ethical textile development and regional production systems. Textile Lab brings together designers and farmers with the goal of creating products that have social and environmental value. Laura also teaches at Parsons School of Design.

Alyssa Schwann is an Assistant Professor in the Faculty of Architecture, Environmental Design program and Department of Landscape Architecture, University of Manitoba, Winnipeg, MB, Canada. She is co-director of the landscape architecture practice, Atelier Anonymous, which studies the convergence of design, landscape conservation, ecology and art.

Avery Shaw is an interdisciplinary designer. Using a process-led approach, her work evolves through different materials and mediums exploring themes of product design in the home, empathy towards objects, and our concepts of the natural environment—how our views of the world influence our understanding, our habits and our space.

Else Skjold is Associate Professor in design and sustainability at the Kolding School of Design (DK). Her research is based in design management studies, in the cross-section between fashion studies, design research and business development. Her departure is the so-called wardrobe method, with focus on the interplay between use, identity-making and clothing.

Huhana Smith, Head of Whiti o Rehua (School of Art) at Massey University, is a visual artist, curator and researcher who engages in major environmental, trans-disciplinary, kaupapa Māori and action-research projects. She actively encourages the use of art and design's visual systems, through exhibition, to expand how solutions might integrate complex issues and make them more accessible for local communities.

Louise St. Pierre researches at the intersection of contemplative practice, environmental activism and design's relationship with nature. She is associate professor at Emily Carr University, Canada and is coordinator of the Emily Carr DESIS lab (Design of Social Innovation for Sustainability).

Paola Sturla architetto (PoliMi) and paesaggista (MLA Harvard), is PhD candidate in Urban Planning Design and Policy at Politecnico di Milano. Before entering Academia, Paola has been practicing for several years as landscape design discipline leader at One-Works, a Milano based engineering company with core expertise in infrastructure design.

Mathilda Tham's work sits in a creative, feminist, activist space between design, futures studies and sustainability. She is Professor of Design, Linnaeus University, Sweden and metadesign researcher, Goldsmiths, University of London.

Jana Thierfelder studied Visual Communication, Transdisciplinarity in the Arts and Research on the Arts. Her research interests cover the intersections between design research, science and technology studies, and social anthropology, investigating the cognitive as well as epistemic role of visual notation techniques within scientific research.

Clara Vankerschaver is a textile designer and researcher. Her practice involves immaterial heritage, anthropology and ecology to formulate tangible ideas about a more durable society. Currently she is working on the project 'Textile as a vessel for tacit knowledge' supported by the research fund of KASK, HoGent.

Markus Wernli is an artisan, researcher, and teacher, dedicated to socially integrative, ecologically engaged practice. He is investigating the dynamics of relational motility in our World of Eaters, where all life forms are eating what is feeding on them—thereby learning to cultivate homemaking capabilities for interexistent flourishing.

Index

C

G

H

I

J

K

L

M

N

O

U

V

W

X

Y

Z

… I wanted to let you know that there have been more pollinators this year.
This feels important. Perhaps especially since we have yet to relinquish winter.
I am still in my thermals.